THE BOOK OF NUMBERS

Also by William Hartston

The Psychology of Chess (with Peter Wason)
The Ultimate Irrelevant Encyclopaedia (with Jill Dawson)
Chess: The Making of the Musical
The Drunken Goldfish
How Was It For You, Professor?
Short v Kasparov, 1993
Teach Yourself Chess
Teach Yourself Better Chess
The Guinness Book of Chess Grandmasters
Odd Dates Only

THE BOOK OF NUMBERS

The Ultimate Compendium of Facts About Figures

WILLIAM HARTSTON

First published in Great Britain in 1997 by
Richard Cohen Books
This second edition published in 2000
by Metro Books, an imprint of
Metro Publishing Limited,
19 Gerrard Street, London W1V 7LA

British Library Cataloguing in Publication Data: a CIP record of this book is available on request from the British Library.
ISBN 1 900512 93 9
10 9 8 7 6 5 4 3 2 1

Typeset by MATS, Southend-on-Sea, Essex
Printed in Britain by CPD Group, Wales

CONTENTS

Acknowledgements

I should like to thank all of my friends and colleagues who have helped my researches by supplying me with their cast-off numbers. The reference works I have consulted are too numerous to mention, but I owe a particular debt of gratitude to the various encyclopaedias on CD-ROM that make the search for numerical information so much easier than it once was. There is a brief bibliography at the end of this book of the works I found most useful, and which the reader in search of more information might consult.

I am grateful to all those who have pointed out errors in the first edition of this book and made suggestions for additions that have been incorporated into this edition. I am particularly indebted to David Singmaster for his many helpful comments.

Preface

This is a book about numbers – to be more exact, the positive integers from one to 5,671,350,435,227. The journey stops at all stations until we are well past the boiling point of water on the Fahrenheit scale, then proceeds at an ever-increasing rate towards its destination. This is not, primarily, a history of numbers or of counting, though anyone interested in those topics will find a good deal of relevant material in the pages that follow. It is not about the mathematical theory of numbers, though many arithmetical curiosities will be found here also. Neither is this just a compilation of number-related lists, though you will find many of those here too. There is also plenty of information about numbers in films, numbers in literature, numbers in song, numbers in nature, indeed, numbers wherever you might encounter them.

This is, in short, a book of numbers unlike any other book of numbers. What I have tried to do is assemble a collection of number-related information and arrange it in numerical order. As a work of reference, it is designed to answer any question beginning with the words 'How many . . . ?', particularly questions that you would never have thought of asking in the first place.

You will find some information on every number from 1 to 216, after which the entries become more sporadic. I have included an introduction to each of the numbers from 1 to 100, to give a guide to the character of the number and particular facts or superstitions that may be related to it. After 100, with a few, too important to overlook, exceptions, you are generally left on your own to draw whatever conclusions you choose from the facts supplied.

I have resisted the temptation to give systematic coverage to any particular sphere, preferring simply to collect and collate information as I bumped into it, or it into me. This accounts for the totally unsystematic policy regarding units of measurement. You will find miles jostling alongside kilometres, pounds next to kilograms, Celsius on equal footing with Fahrenheit. My only rule has been to keep all measurements in the same units as I first encountered them.

You may open the book wherever you like and start reading. Indeed, the material may perhaps be best enjoyed in the same manner in which it has been collected: totally haphazardly. The only signposts are an occasional → which I have inserted to point towards a related entry under another number.

Zero

'I got plenty o' nothing'
George Gershwin, *Porgy and Bess*, 1934

There were no people injured by tea-cosies in Britain in 1994 (though three had been treated in hospital in the previous year as a result of tea-cosy accidents). By 1997, however, the tea-cosy injury rate had risen again to two, showing what a hazardous life the dedicated tea-drinker leads. All UK home injury figures in this book come from the annual survey published by the Office for National Statistics and represent the number of people treated for each type of injury in 18 hospitals around the country. The total number of tea-cosy injuries may thus be assumed to be far greater, though complete national records are sadly not available.

Zero is also the number of times the word 'Bible' occurs in the works of Shakespeare.

Those items, however, do not really concern us. For this book is about the positive integers, from 1 to over five trillion.

1

'One is one and all alone and ever more shall be so.'
From *The Dilly Song*, also known as *Green Grow the Rushes-O*

The ancient Greeks did not consider one to be a number at all. Euclid had defined 'number' as an 'aggregate of units', so one is not so much a number itself as the mother of all numbers. Even more oddly, however, they regarded the number one as both odd and even – perhaps simply to support its unique (from the Latin, *unus*; one) claim to be both male and female and thus the number from which all others sprang.

In religious mysticism, one represents God, while in numerology one is associated with the Sun, and with everything positive and original. Linguistically, one-ness is signified by the prefixes mono- or uni- (including the words unity, union and universe), while even the words 'a', 'an' and 'alone' are etymologically close relatives of 'one'.

One is also the number of:

• Butler schools in the US
• cases of acute poliomyelitis in the UK in 1995
• centimetres each hair on a human head grows each month
• dead heats in the Oxford-Cambridge boat race
• elephants in Alaska in 1995 (the date when the Alaskans changed their laws relating to the keeping of elephants specifically to allow one ex-circus elephant to stay there)
• eyes on a Cyclops
• heliports in Algeria
• hiccups in the works of Shakespeare (uttered, appropriately enough, by Sir Toby Belch in *Twelfth Night*)
• horns on a unicorn
• minutes each year the average Haitian spends making international phone calls
• paintings sold by Vincent van Gogh in his lifetime
• public telephones in Kabul
• recorded accidents in the home in the UK in 1994 involving whistles
• shove ha'penny boards in Swaziland
• small steps for a man ('One giant leap for mankind')
• times the word 'girl' is in the King James Bible: 'And they have cast lots for my people; and have given a boy for an harlot, and sold a girl for wine, that they might drink.' (Joel 3:3)
• shots under par for a birdie at golf
• US presidents born under the sign of Gemini (George Bush)
• wheels on a unicycle
• years' marriage for a cotton anniversary

The word 'one' has occurred in the titles of at least 500 films of which the following are of particular note:

- *One A.M.* (1914): a drunken Charlie Chaplin spends twenty minutes trying to climb the stairs to bed
- *Johnny One-Eye* (1950): Pat O'Brien in a Damon Runyon story
- *Number One* (1969): Charlton Heston as an ageing football player
- *One Day in the Life of Ivan Denisovich* (1971): Tom Courtenay in the title role of this Solzhenitsyn adaptation
- *One is a Lonely Number* (1972): drama based on a story *The Good Humor Man* by Rebecca Morris
- *The Magnificent One* (1974): Jean-Paul Belmondo in a James Bond spoof
- *1 = 2?* (1975): French comedy
- *One Flew Over the Cuckoo's Nest* (1975): Jack Nicholson in the multiple-Oscar-winning film of Ken Kesey's book
- *Capricorn One* (1978): O. J. Simpson turns up in this tale of NASA faking a Mars landing
- *One Trick Pony* (1980): written and scored by Paul Simon
- *Duet for One* (1986): Julie Andrews and Max von Sydow in a drama about a violinist

And before they go astray, we must mention four more:

- *One of Our Aircraft is Missing* (1941)
- *One of Our Spies is Missing* (1966)
- *One of Our Dinosaurs is Missing* (1975)
- *One of My Wives is Missing* (1976)

The number one also occurs in a wide range of idiomatic expressions including: one-armed bandit, one foot in the grave, one-horse town, one-night stand, one-track mind, one-way street, one-way ticket, one-man band and one fell swoop.

2

'There are two sides to every question.'
Protagoras, Greek philosopher and sophist of the fifth century BC, believed to have been the first person to study and write on grammar.

While the Greeks objected with intensity to thinking of one as a number, they had considerable reservations about two as well. It was a perfectly good number in many respects, and useful for counting and multiplying in a way that one was not, but they still had doubts. Two could be said to have a beginning and end, but no middle. The geometers in particular felt that two did not fulfil the required standards: you can have a perfectly good three-sided figure, but a two-sided one falls flat.

Two, in religious mysticism, is the number of disunity and the number of creation – because two represents the splitting apart of divine unity. It is, perhaps, no coincidence that both the Old Testament and the Koran begin, in their original languages, with the second letter of the alphabet: *b'reshit* (In the beginning) and *bismillah* (In the name of God).

Linguistically, two is associated with the prefixes bi- (Latin) and di- (Greek), though sloppiness creeps in with words such as bicycle (Latin two, Greek wheels) and bigamy (Latin two, Greek marriages) which ought to be dicycle and digamy. A more hidden two is found in the word 'diploma', which is a certificate folded in two. Also several words beginning tw- reflect the number two: a twig is where a branch splits into two; twine is a rope formed by plaiting two threads together. One word that has no etymological connection with two-ness is 'bikini', named after the Pacific atoll where nuclear tests were held. The jocular suggestion of a swimsuit blown into two parts might never have become accepted but for the suitability of the apparent bi- prefix. It even led to the later words 'monokini' and 'trikini' for one- and three-piece swimsuits.

Two is also the number of:

• Barbie dolls sold somewhere on earth every second

• bottles in a magnum
• Gentlemen of Verona
• kilograms minimum weight for a man's discus
• nickels in a dime
• pipes in a tun
• place names in the US that include the word 'sex': Sex Peak and Sex Peak Lookout, both in Montana
• places on earth named 'Hell', one in Norway, the other in the Cayman Islands
• recorded accidents in the home in the UK in 1994 involving kitchen scales
• sets of false teeth found on British beaches during the Beachwatch survey in 1998
• shots under par for an eagle at golf
• teaspoons in a dessertspoon, and dessertspoons in a tablespoon though British spoons and American spoons appear to differ in size: in North America a teaspoon is officially 4.833 millilitres and equals one third of a tablespoon; in Britain a teaspoon is 3.625 millilitres and is a quarter of a tablespoon
• times Princess Diana uttered the name 'Charles' during her hour-long Panorama interview on television in 1995
• times the word 'cheese' is in the King James Bible
• turtle doves my true love gave to me
• wrongs that don't make a right
• years' marriage for a paper anniversary

The word 'two' has turned up in the titles of some 300 films including the following:

• *The Man with Two Faces* (1934): Edward G. Robinson in a crime mystery
• *Breakfast for Two* (1937): Barbara Stanwyck comedy romance
• *Two Sisters from Boston* (1946): musical with Jimmy Durante
• *The Two Mrs Carrolls* (1947): Humphrey Bogart and Barbara Stanwyck

in a tale of a serial wife-killing artist
• *Eagle With Two Heads* (1948): Jean Cocteau romance
• *A Kid for Two Farthings* (1955): David Kossoff stars in an adaptation of Wolf Mankowicz's novel
• *Two Men and a Wardrobe* (1958): a slapstick short made by Roman Polanski while still a student that helped establish his reputation
• *Only Two Can Play* (1962): loose adaptation of Kingsley Amis's *That Uncertain Feeling*, with Peter Sellers
• *Two for the Seesaw* (1962): Shirley MacLaine and Robert Mitchum in a comedy romance
• *Flat Two* (1962): Edgar Wallace mystery
• *Two for the Road* (1967): Albert Finney and Audrey Hepburn in a story by Frederick Raphael
• *The Magnificent Two* (1967): Morecambe and Wise comedy
• *Two Mules for Sister Sara* (1970): Clint Eastwood and Shirley MacLaine in nineteenth-century Mexico
• *The Incredible Two-Headed Transplant* (1971): low-budget horror
• *Two English Girls* (1972): François Truffaut adaptation of a novel by Henri-Pierre Roche
• *Two Against the Law* (1973): Alain Delon and Jean Gabin
• *Number Two* (1975): experimental film about a constipated wife and impotent husband
• *2 Catch 2* (1979): drama about a gambler's high-risk attempt to recoup his losses
• *Chapter Two* (1979): James Caan and Marsha Mason in adaptation of Neil Simon's stage hit
• *The Man With Two Brains* (1983): Steve Martin and Sissy Spacek in sci-fi spoof
• *A Zed and Two Noughts* (1985): decaying matter beautifully filmed by Peter Greenaway

3

'What I tell you three times is true.'
Lewis Carroll, *The Hunting of the Snark*, 1876

Good things come in threes: animal, vegetable and mineral; solid, liquid and gas; gold, silver and bronze medals; ego, id and superego; reading, writing and arithmetic; God the Father, Son and Holy Ghost. Three is the first number that all Greeks were able to agree was worthy of the term 'number'. As Aristotle pointed out, it is the lowest number to which the term 'all' can properly be applied. And three repairs the damage that two did, in splitting things apart.

Numerically, three is the second triangular number: 3 = 1+2; the next triangular number is 6 which is 1+2+3, then 10 (1+2+3+4) and so on. Imagine a single point with two points below it, three below them, four below them and so on, forming an ever-growing triangle and you will see where the name comes from. Anyway, in 1796, Gauss, at the age of nineteen, proved that every number is the sum of at most three triangular numbers.

Linguistic triples begin with tri- in both Greek and Latin derivations. Three of the less obvious are 'trivial' – meaning three roads, indicative of the sort of talk you get at crossroads; 'tripos' – the name given to examinations at the universities of Oxford and Cambridge, but the word refers not to the three years of an undergraduate course or three parts of an examination but to the three legs of the stool on which the examiner sat; and 'trilogy' – which ought really to mean three words, not three books.

Three is also the number of:

- baths taken by Louis XIV in his entire lifetime
- blind mice
- days Jonah spent in the belly of the whale
- dimensions of the physical world

• Fates: Lachesis (who determined the fate), Clotho (spinner of thread of life), Atropos (who cut the thread)
• feet in a yard
• French hens my true love gave to me
• Friday the thirteenths in 1998, the highest number possible in a calendar year
• Furies: Alecto (the unresting), Megaera (the jealous), Tisiphone (the avenger)
• garden hoes in the traditional dowry for a wife in Rwanda
• Graces (Aglaia, Thalia, Euphrosyne – the daughters of Zeus and Eurynome)
• highest number in the language of the Yancos tribe of the Amazon: their word for three is 'Poettarrarorincoaroac'
• highest number of overcoats found in the stomach of a single shark
• illegitimate children of Friedrich August I of Saxony (King Augustus II of Poland) who became field marshals
• injuries recorded in Britain in 1997 caused by accidents involving pencil sharpeners
• little maids from school in Gilbert and Sullivan's *Mikado*: (Yum-Yum, Peep-Bo, Pitti-Sing)
• little pigs (and one big bad wolf)
• Musketeers (Athos, Porthos, Aramis)
• oranges in *The Love for Three Oranges* (opera by Prokofiev)
• ounces of protein eaten per head daily in UK
• people per square kilometre in Canada
• people who died when Cachi the poodle fell from a balcony in Buenos Aires in 1988: one woman was killed by the dog falling on her head, another was so startled by seeing it that she stepped backwards into the path of a bus, and a third person died of a heart attack
• primary colours: red, yellow and blue
• recorded accidents in the home in the UK in 1994 involving weights for kitchen scales
• sheets in the wind, an old expression for drunkenness, alluding to the

unsteadiness of a sailing ship whose three sails have all become unfastened in a high wind
• Stooges (Larry, Moe and Curly)
• Strikes and You're Out
• teeth of his own that Stalin had left at his death
• theological virtues: Faith, Hope and Charity
• times Philip IV of Spain is said to have smiled in his life
• under par for an albatross at golf
• wheels on a tricycle
• Wise Men (Gaspar, Melchior, Balthasar)
• Wise Monkeys (the Apes of Nikko: Mizaru, Mikazaru, Mazaru – see, hear and speak no evil)
• witches in *Macbeth*
• words in English still taking the old -en in the plural: oxen, brethren, children
• years' marriage for a leather anniversary

Among the many films featuring the number three are the following:
• *The Three Ages* (1923): Buster Keaton silent comedy
• *These Three* (1936): Merle Oberon in an adaptation of a play by Lillian Hellman
• *Three Faces West* (1940): John Wayne beats the bad guys again
• *The Three Caballeros* (1945): Donald Duck teams up with a parrot and a rooster
• *Three Strangers* (1946): Sidney Greenstreet and Peter Lorre in a tale of gambling and violence
• *The Three Godfathers* (1948): Peter Kyne's *Three Godfathers* was filmed five times, with John Wayne, in this version, the only one of the fifteen Godfathers to survive
• *A Letter to Three Wives* (1949): Kirk Douglas in a tale of suspected infidelity
• *A Letter to Three Husbands* (1950): Emlyn Williams in a tale of

suspected infidelity

• *Soldiers Three* (1951): Stewart Granger, Robert Newton and David Niven in a tale of the British army in India, loosely based on Kipling
• *Three Coins in the Fountain* (1954): double Oscar-winning musical
• *The Three Faces of Eve* (1957): psychodrama that won an Oscar for Joanne Woodward
• *Three Moves to Freedom* (1960): Claire Bloom and Curt Jurgens in an adaptation of Stefan Zweig's *The Royal Game*
• *The 3 Worlds of Gulliver* (1960): adaptation of Jonathan Swift
• *The House of the Three Girls* (1961): Austrian tale of unrequited love based on the life of Schubert
• *The Three Lives of Thomasina* (1963): adaptation of a Paul Gallico novel
• *Three Faces of a Woman* (1965): Michelangelo Antonioni directs Richard Harris
• *Three* (1969): Charlotte Rampling in a love triangle
• *Three Into Two Won't Go* (1969): Rod Steiger falls for Judy Geeson
• *Three Days of the Condor* (1975): Max von Sydow and Faye Dunaway in an adaptation of James Grady's book, *Six Days of the Condor*
• *Three Women* (1977): Robert Altman directs Shelley Duvall and Sissy Spacek in psychodrama
• Agatha Christie's *Murder in Three Acts* (1986): Peter Ustinov finds the murderer at Tony Curtis's parties
• *Three Men and a Baby* (1987): Ted Danson, Steve Guttenberg and Tom Selleck babysit
• *Three Men and a Cradle* (1985): the French original of the above
• *Three Men and a Little Lady* (1990): the Baby grows up

Among the novels with three in their titles, the best known are:

• *The Three Musketeers* by Alexandre Dumas
• *Three Men in a Boat* by Jerome K. Jerome

Best play:

- *The Three Sisters* by Chekhov

Operas:

- *The Love of Three Oranges* by Prokofiev
- *The Threepenny Opera* by Kurt Weill and Berthold Brecht

4

'Four legs good, two legs bad.'
George Orwell, *Animal Farm*, 1945

Four is the only number equal to the number of letters in its English word. The same applies in German, as *vier* has four letters, but there is no number-word in French that equals its number of letters, though Spanish and Russian have *cinco* (five) and *tri* (three) respectively. The Greeks, of course, did not know this, but they did associate the number four with earthly balance. Earth, air, fire and water were supposedly the elements out of which everything was composed; there were four humours which combined to produce a person's temperament; and north, south, east and west were the four points of the compass from which the Four Winds could blow you to any of the four corners of the earth.

In other cultures, the number four may have less positive overtones. In 1995 Taipei allowed residents to delete '4' from street numbers because it sounds like 'death' in Chinese. For the same reason the previous year they issued car number plates excluding the number 4. Many hospitals in China do not have a fourth floor.

Mathematically, there are a number of important theorems associated with the number four, none more intriguing than the Four-Colour Problem. The question, first posed in the middle of the nineteenth century, concerned how many colours you need to colour a map with the sole criterion being that no two areas (countries or counties, for example) sharing a common stretch of border may be the same colour. Many trials failed to produce any map, however complicated, that required more than four colours, yet for more than a century nobody could prove that four was always sufficient. The problem was finally solved in 1976 when the general theorem was proved except for a large number of possibly anomalous cases, and a computer was set to work for several weeks eliminating all the possible exceptions.

Linguistically, four is the number where the sequence once, twice, thrice comes to a stop. Foursomes may be indicated by the prefixes quadr- (Latin) or tetra- (Greek). Purists, therefore, were not surprised that when record companies in the 1960s launched 'quadraphonic' recordings the new technology met with an unreceptive market. It should, to maintain linguistic consistency, have been called 'quadrasonic' or 'tetraphonic'.

Rude words traditionally have four letters and, according to the evidence of *Chambers Dictionary*, this reputation is well justified. Of the words listed in the dictionary as 'vulg' there are 22 with four letters compared with five of three letters (or six if you include an Americanism), ten of five letters, two of six and only one each of 7, 8, 9, 11 and 12 letters. The expression 'four-letter word', incidentally, dates back only to 1929. It must be mildly confusing for Francophones since the French for 'he let out a four-letter word' is, according to *Collins Dictionary*, 'il a sorti le mot de cinq lettres'.

Four is also the number of:

- balls in a croquet set (blue, black, red, yellow)
- bottles in a jeroboam
- calling birds my true love gave me on the fourth day of Christmas
- cups of tea drunk each day by the average Briton
- dimensions of Einstein's space-time continuum
- eggs eaten per week per capita in the UK
- Evangelists – Matthew, Mark, Luke, John

• fatal accidents in British homes caused by stationery/writing equipment in 1992
• fingers on each of Mickey Mouse's hands
• freedoms mentioned by F. D. Roosevelt on 6 January 1941: freedom of speech and expression; freedom of religion; freedom from want; freedom from fear
• frogs eaten each year per capita in France
• gates of Roman London, on the sites later known as Dourgate, Aldgate, Aldersgate and Ludgate. In the seventeenth century, the walled city could also be entered by Bishopsgate, Moorgate, Cripplegate and Newgate. As well as these, there were the Postern Gate on Tower Hill and Bridge Gate, but neither of those was an entry point through the wall itself
• grams of salt in a litre of sweat
• Honorary degrees held by Sammy Davis jnr
• Horsemen of the Apocalypse: War, Famine, Pestilence, Death
• humours: black bile, yellow bile, phlegm, blood
• inches in a hand (when used as a measure of height for horses)
• mistakes identified on the map of Europe on the Italian 1000 lire coin issued in 1998
• months an oyster can survive out of water
• ounces of fat eaten per head daily in the UK
• players in a game of whist or bridge – while whist takes its name, according to Cotton's *Complete Gamester* of 1680, from the 'silence that is to be observed in the play', the etymology of the name of the game of bridge (which dates back to around 1870) is unascertained, but it may be connected with a Slavonic word, *birritch*, meaning 'without trumps'
• players on a polo team
• quarts in a gallon
• recorded accidents in the home in the UK in 1997 involving chopsticks
• recorded accidents in the home in the UK in 1994 involving Christmas tree lights

• riots at golf events in the US between 1960 and 1972
• roods in an acre
• times Norway has scored 'nul points' in the Eurovision Song Contest
• times the word 'Queen' occurs in the first verse of the National Anthem
• toes on the front paw of the abominable snowman according to an alleged sighting in 1953
• years' marriage for an iron (or some say fruit/flowers) anniversary

There are over a hundred films with the word or number four in the title, including the following:
• *The Four Horsemen of the Apocalypse* (1921): with Rudolph Valentino doing the tango
• *Four Frightened People* (1934): Claudette Colbert and Herbert Marshall in a Cecil B. de Mille drama
• *Four's a Crowd* (1938): comedy with Errol Flynn and Olivia de Havilland
• *The Four Feathers* (1939): love and cowardice in the desert with Sir Ralph Richardson, based on the book by A. E. W. Mason
• *The Four Just Men* (1939): based on Edgar Wallace's book
• *Adam Had Four Sons* (1941): Robert Shaw and Ingrid Bergman drama
• *4-D Man* (1959): hokum in which scientist enters fourth dimension and goes mad
• *Four Days in November* (1964): Kennedy assassination documentary
• *The Four Musketeers* (1975): Michael York's d'Artagnan is upgraded for this sequel
• *The Four Seasons* (1982): Alan Alda starred and directed
• *Mishima: A Life in Four Chapters* (1985): biography of the writer Yukio Mishima who committed suicide after failing in a right-wing coup
• *Adventures of Sherlock Holmes – The Sign of Four* (1985): Jeremy Brett as Holmes
• *Four Weddings and a Funeral* (1994): very British romantic comedy

5

'Five is the human soul. As man is a mixture of good and evil, so is Five the first number made from both even and uneven.'
Schiller: *Fünf ist des Menschen Seele. Wie der Mensch aus Gutem und Bösem ist gemischt, so ist die Fünfe die ersten Zahl aus Grad' und Ungerade* – from *Piccolomini*, 1799

Five is an odd number, in both senses of the word. We have five fingers on each hand, five toes on a foot and five senses; a starfish has five legs and flowers frequently have five petals, yet true five-fold symmetry occurs rarely in nature. The Pythagoreans, of course, had the answer: as the sum of two (female) and three (male), five was the number of love and marriage. Pythagoras's Theorem itself was sometimes known as the 'Theorem of the Bride', in recognition of the smallest right-angled triangle with sides of integral length: three (male) and four (female) containing the right-angle produced the offspring of a hypotenuse of length five. Alternatively, you could add one (the number of divine creation) to four (earthly balance) to get the answer five (the magic of reproduction) – which may help to explain the supposed power of the pentagram in magic. To complete the human connection, five is the number of limbs plus one for the body (or head).

Mathematically, five is the number of Platonic solids – the solid bodies all of whose faces are congruent regular polygons: the tetrahedron (whose four faces are all identical equilateral triangles); cube (six squares); octahedron (eight triangles); dodecahedron (twelve pentagons); and icosahedron (twenty equilateral triangles). Plato knew about all of them and Euclid proved that there were no more than these five.

Since we are all told that our number system uses the base of ten because we have ten fingers for counting on, it might seem legitimate to wonder why we did not evolve a system of numbers to base five (perhaps with the intention of leaving the other hand free for doing something more useful than mere counting). In fact, the only language known to have a counting system based on five rather than ten is Saraveca, one of the Arawakan languages of South America.

Linguistically, sets of five are denoted by the prefix penta- (Greek) or quin- (Latin). Quintets, quintuplets and pentathlons may be quintessentially familiar, but most of us will have forgotten that 'quintessence' – the purest essential nature of something – is the fifth essence, after earth, air, fire and water. The throat inflammation known as quinsy, however, has nothing at all to do with the number five, but comes from the Latin *quinancia*, which derives from the Greek *kynanche* from *kyon*, dog + *anchein*, strangle, through whether it is supposed to make you feel or sound like a dog being strangled is lost to etymological history.

Five is also the number of:

- baseball gloves that can be made from one cow
- Books of Moses comprising the Torah (or Pentateuch)
- conductors needed for a performance of Henry Brant's *Antiphony 1*
- fluid ounces in a gill
- gold rings my true love gave me
- Great Lakes of North America: Superior, Michigan, Huron, Erie and Ontario, of which Lake Michigan is the only one entirely in the United States
- inches length of President Clinton's erect penis according to Paula Jones
- kilograms of sweets other than chocolate eaten per person per year in the UK
- magic beans in Jack and the Beanstalk
- minutes it takes a good vet to castrate a cat
- paintings by Picasso in the top ten most expensive sold at auction (the others comprise three by van Gogh and one each from Renoir and Jacopo da Carucci)
- pints capacity of an average dog's stomach
- recorded accidents in the home in the UK in 1994 involving strainers or sieves

• recorded accidents in the home involving pillowcases in 1997 in the UK
• Rivers of Hades: Acheron (river of woe); Cocytus (lamentation); Lethe (oblivion); Phlegethon (fire); Styx (hate)
• times the average British woman thinks about sex each day (according to a 1997 survey by *Cosmopolitan* magazine)
• victories over enemy aircraft to qualify to become an 'ace'
• ways the letter 'f' may be pronounced in Icelandic
• women named Mary in the New Testament
• years it took Marva Drew of Waterloo, Iowa to type all numbers from 1 to 1,000,000
• years' marriage for a wood anniversary

Finally, we should mention a useful old saying: Spring has arrived when you can cover five daisies with your foot.

Over a hundred films have 'five' in the title including:
• *The $5 Counterfeiting Plot* (1914): a six-minute silent film
• *Five Star Final* (1931): Edward G. Robinson and Boris Karloff in a tale of journalistic immorality
• *Five Little Peppers and How they Grew* (1939): children's story from the book by Margaret Sidney
• *The Beast with Five Fingers* (1946): horror story from the book of the same name by William Fryer Harvey
• *Five* (1951): only five people survive a nuclear holocaust
• *The Sheep Has Five Legs* (1954): classic French comedy with Fernandel in six roles
• *Five to One* (1963): Dawn Adams and John Thaw in an Edgar Wallace crime story
• *Five-Card Stud* (1968): Robert Mitchum and Dean Martin in a tale of poker and murder
• *Five Easy Pieces* (1970): Jack Nicholson and Karen Black in a romantic drama

• *Slaughterhouse Five* (1972): adaptation of Kurt Vonnegut's satirical novel
• *Five Fingers of Death* (1973): martial arts epic which some say started the western Kung Fu craze
• *Five Miles to Midnight* (1980): adventure with Joan Collins, Roger Moore and Tony Curtis
• *Dragon Lee vs the Five Brothers* (1981): martial arts
• *Five Days One Summer* (1982): romance and suspense with Sean Connery
• *Come Back to the Five and Dime, Jimmy Dean, Jimmy Dean* (1982): Karen Black in a sex-change drama

6

'Shakespeare never had six lines together without a fault. Perhaps you may find seven, but this does not refute my general assertion.'
Samuel Johnson, 1769

The number six has been imbued with great significance ever since God took that number of days to create the world. Mathematically, six is a perfect number (one that is equal to the sum of its proper divisors: 6 = 1+2+3), and St Augustine was among those who argued that God chose six days for the Creation precisely because six was perfect. Nature also finds the number six useful in its own creative processes, as may be seen in the hexagonal structure of the snowflake and the bees' honeycomb, and in the benzene ring, C_6H_6, the idea of which is said to have come to the German chemist Friedrich Kekulé von Stradonitz in a dream.

Mathematically, six is a triangular number (1+2+3) and a factorial (3×2×1) as well as being perfect. Here's an old trick involving the number six: think of a non-zero number (9, for example), and multiply it by three (27). Count backwards two steps from the answer (27, 26, 25) and add all three numbers together (27+26+25 = 78). Add together the digits of your answer (7+8 = 15) and do so again until you reach a single digit (1+5 = 6). The answer will always be six.

Linguistically, sixes are signified by hexa- (Greek) or sexa- (Latin). A semester was originally a term of six (Latin: *sex*) months (*menses*), and even the Spanish *siesta* began as a rest at the sixth hour of the day (around noon, if you get up early enough). A sextant is a device designed to measure angles up to 60°, which is a sixth of a circle.

Six is also the number of:

• bottles in a rehoboam
• characters in search of an author in the Pirandello play
• chicken sandwiches sold every second in Britain
• deaths caused by the Great Fire of London
• feet in a fathom
• geese-a-laying that my true love gave to me
• international airports in Pennsylvania
• ligulae (= 1/50 pint) in an acetabulum – a word the Romans used for a vinegar pot, which is now the term for the socket of the hip joint
• men hanged for sodomy in England and Wales in 1806
• men shaved in one minute by champion barber Robert Hardie in 1909
• months in the life of the average American spent waiting for red lights to change
• pints held by the average ten-gallon hat
• players on an ice-hockey team
• points for a Q in Portuguese Scrabble
• radios in the average US household
• ratio of the earth's gravity to the moon's
• recorded accidents in the home in the UK in 1994 involving bidets
• sides on a snowflake
• toes Charles VIII of France had on one foot (having one more than the standard number of digits on hand or foot is known as hexadactylism; the term covering any number of extra digits is polydactylism. Anne

Boleyn is also believed to have been hexadactylic, having six fingers on one hand). The West Indian cricketer Garry Sobers was born with a supernumerary finger on each hand, but this was corrected by an early operation
• ways Shakespeare spelt 'Shakespeare'
• wives of Henry VIII
• years' marriage for a sugar anniversary

There are fewer than a hundred films with 'six' in the title. Here are some of them:
• *The Secret Six* (1931): crime mystery with Clark Gable and Jean Harlow
• *Six Bridges To Cross* (1955): crime drama with Tony Curtis and Sal Mineo – from a book by Joseph F. Dinneen called *They Stole $2,500,000 and Got Away With It*
• *Two and Two Make Six* (1962): comedy drama
• *With Six You Get Egg Roll* (1968): Doris Day comedy
• *6 Rms Riv Vu* (1975): Alan Alda and Carol Burnett comedy
• *Six Weeks* (1982): love story with Mary Tyler Moore and Dudley Moore
• *Six Degrees of Separation* (1993): Sidney Poitier true-life drama
• *06* (1994): dirty Dutch film – 06 is the phone code for sex lines in the Netherlands

Moving slightly ahead of six, we should mention that 'half-past six' is Singaporean English for not completely sane.

And while we are midway between six and seven, we should take the opportunity to discuss the origin of the phrase 'at sixes and sevens' meaning totally confused. The most likely explanation connects the phrase with an old dice game called 'hazard' at which a player could face a difficult quandary through not knowing whether to be on six or seven. A less convincing explanation links it to the struggle between the Merchant Taylors and the Skinners Companies for sixth and seventh place in the table of precedence of the London Guilds. The problem with that explanation, however, is that the earliest use of the expression probably predates the dispute between the Guilds.

7

'Wisdom hath builded her house, she hath hewn out her seven pillars.'
Proverbs 9:1

Adding the day of rest to the six days of creation, we arrive at the seven-day week and the most magically mystical number of all. In ancient Egypt there were seven paths to heaven. The ancient Babylonians had seven levels in their ziggurat, or step-pyramid, and their tree of life had seven branches, as also did the original version of the symbolic Jewish candlestick, the menorah. The ancient Chinese saw seven as the number governing female life; milk teeth arrive at seven months and fall out at seven years; puberty arrives at twice seven years and the menopause at seven times seven. The menorah, with its central shaft and three branches on each side, made of pure gold by Bezaleel, is described in Exodus, chapter 37. It fits well alongside many other sevens in the Old Testament, from the seven times Cain's murder will be avenged to the seven days Noah's dove spent away from the ark, the seven steps leading to Solomon's temple (which took seven years to build), the seven locks of Samson, seven sneezes of a child raised from the dead by Elijah (2 Kings 4:35), and many others. The New Testament carries on the worship of heptads with the seven gifts of the holy spirit, seven last words of Christ on the cross and seven seals and seven trumpets in the Book of Revelation. By the time Shakespeare wrote of the seven ages of man the idea had already been firmly established for many centuries.

No wonder, then, that seven is associated with good luck, even in modern times. In Japan, for example, at seven minutes past seven, on the seventh day of the seventh month of the seventh year of the Japanese emperor's reign (1995), 17 runners ran 7,777 metres round the imperial palace, and all because the number seven is considered lucky. And that is also why we drink 7-Up and fly in Boeing aeroplanes known by a single digit sandwiched between two sevens.

Despite this generally positive view of seven, it is also the number of the Deadly Sins: pride, envy, wrath, sloth, avarice, gluttony and lust. When Pope Gregory I (later known as St Gregory the Great) drew up this list around AD 600, sloth, in its present meaning of idleness, was not among them. In its place was accidie - a sort of boredom or torpor that was seen as a denial of the magnificence of creation and God's work. George Bernard Shaw, however, had his own list of seven deadly sins: food, clothing, firing, rent, taxes, respectability and children.

Psychologically, seven is approximately the number of different things we can hold in our short-term memories at the same time. This idea was first expressed in 1956 in a paper by G. A. Miller entitled: 'The magical number seven plus or minus two: some limits on our capacity for processing information' (*Psychological Review*, vol. 63, pp. 81–97). Subsequent research has confirmed that our ability to remember strings of words, numbers or other items is generally limited to strings of between five and nine, depending on the complexity of each item, unless we have specially trained ourselves for a specific task such as remembering playing cards or series of digits.

Linguistically, sevens usually come in the two varieties of Greek, hepta- and Latin, septem-. Particular note should be taken of the word 'septemvious' meaning in seven different directions, of which the *Oxford English Dictionary* supplies only one usage, from 1861. Slightly more common, particularly in Oxford University, is the word 'hebdomadal' meaning once every seven days, hence once a week, which is how often the Hebdomadal Council meets.

Academically, a heptatechnist is a professor of the 'Seven Arts' (also known as the 'free' or 'Liberal' arts), a term first applied in the Middle Ages to a course of seven sciences, dating back to the sixth century, comprising grammar, logic and rhetoric (these three known as the *trivium*) as well as arithmetic, geometry, music, and astronomy (the *quadrivium*).

Seven is also the number of:

- archangels in the earliest references
- colours of the rainbow: red, orange, yellow, green, blue, indigo, violet (though it has been suggested that Isaac Newton added indigo to the list in 1704 only to make the number up to the magical seven)
- Dwarfs encountered by Snow White: Sleepy, Happy, Dopey, Bashful, Grumpy, Sneezy, Doc
- eclipses (lunar + solar) possible in a single year
- grams a human eyeball weighs
- Heavens in the Muslim religion: Silver, Gold, Pearl, White Gold, Silver & Fire, Ruby & Garnet, Divine Light Impossible for Mortal Man to Describe
- Hills of Rome: Aventine, Caelian, Capitoline, Esquiline, Palatine, Quirinal, Viminal
- kings present at the golden wedding celebrations of Queen Elizabeth II in 1997
- length in feet of an average blue whale's penis
- minutes length of a chukkah in polo
- the pH value of pure water (less than seven = acidic, more than seven = alkaline)
- the percentage of methane in a fart
- Pillars of Wisdom of T. E. Lawrence
- players on a netball team
- points for a D in Finnish Scrabble
- recorded accidents in the home in the UK in 1997 involving putty
- recorded accidents in the home in the UK in 1994 involving tape measures
- Seas: Antarctic, Arctic, N. Atlantic, S. Atlantic, Indian, N. Pacific, S. Pacific
- Sisters (in North London)

• Sisters (US Ivy League colleges)
• Sisters (the Pleiades constellation of stars)
• skeins in a hank
• swans-a-swimming given by my true love
• times round the world all the Barbie dolls ever sold would stretch
• Virtues: Faith, Hope, Charity, Fortitude, Justice, Prudence, Temperance
• voyages of Sinbad
• Wonders of the Ancient World: Colossus of Rhodes, Pyramid of Cheops, Hanging Gardens of Babylon, Pharos Lighthouse of Alexandria, Mausoleum at Halicarnassus, Statue of Zeus at Olympia, Temple of Diana at Ephesus
• years of age at which Chopin wrote his first polonaise
• years' marriage for a wool anniversary

The magnificence of 'seven' has made it a popular number to appear on screen. Here are some of the titles:

• *Seven Chances* (1925): Buster Keaton silent movie about a lawyer who will inherit $7m if he marries by 7 p.m.
• *Bluebeard's Seven Wives* (1926): silent film comedy
• *Snow White and the Seven Dwarfs* (1937): Disney classic
• *The Door With Seven Locks* (1940): Edgar Wallace horror story
• *The House of the Seven Gables* (1940): Vincent Price and George Sanders in a faithful adaptation of Nathaniel Hawthorne's novel
• *Madonna of the Seven Moons* (1945): based on a book of the same title by Margery Lawrence
• *Seven Keys to Baldpate* (1947): Jason Robards in the film of the book by Earl Derr Biggers
• *The Seven Samurai* (1954): Akira Kurosawa's classic tale starring Toshiro Mifune
• *Seven Brides for Seven Brothers* (1954): Oscar-winning musical with Howard Keel
• *The Seven Year Itch* (1955): Marilyn Monroe film from the play by

George Axelrod
• *The 7th Voyage of Sinbad* (1958): fantasy adventure
• *The Seven Hills of Rome* (1958): musical with Mario Lanza
• *The House of the Seven Hawks* (1959): Donald Wolfit and David Kossoff in a tale of Nazi treasure
• *The Magnificent Seven* (1960): western remake of Kurosawa's *The Seven Samurai*
• *Robin and the Seven Hoods* (1964): musical gangster comedy with Bing Crosby, Frank Sinatra and Dean Martin
• *Seven Days in May* (1964): from the book of the same title by Fletcher Knebel and Charles W. Bailey
• *Return of the Magnificent Seven* (1966): the first sequel, though only Yul Brynner returns
• *The Duck Rings at Half Past Seven* (1969): German comedy
• *Guns of the Magnificent Seven* (1969): sequel
• *The Magnificent Seven Deadly Sins* (1971): Spike Milligan and Harry Secombe comedy
• *The Magnificent Seven Ride* (1972): the final sequel
• *The Seven Brothers Meet Dracula* (1974): martial arts vampire movie with Peter Cushing
• *The Seven Percent Solution* (1976): Alan Arkin as Sigmund Freud, Nicol Williamson as Holmes and Sir Laurence Olivier as Moriarty in an adaptation of Nicholas Meyer's book
• *Dracula and the Seven Golden Vampires* (1978): another martial arts vampire film with Peter Cushing
• *Seven* (1979): gunslinger is paid $7m by the US government to deal with some gangsters
• *Seven Sixgunners* (1987): western adapted from the book by Nelson Nye
Seven (1995): Brad Pitt and Gwyneth Paltrow in a serial-killer mystery horror based on the seven deadly sins

Finally, let us not forget the name by which the secret ingredient of Coca-Cola is known: 7X

8

'And on the eighth day, the flesh of his foreskin shall be circumcised.'
Leviticus 12:3

Eight is an auspicious number through whichever religion you look at it. The Muslims believe in eight paradises (and seven hells): Christ, in his Sermon on the Mount, mentioned eight beatitudes; Buddhism teaches the noble eightfold path of enlightenment (right understanding, right aspiration, right speech, right conduct, right means of livelihood, right endeavour, right mindfulness and right contemplation); and Norse mythology gave Odin an eight-legged horse called Sleipnir and a ring, Draupnir, from which eight rings of equal value dropped off every nine days.

Mathematically, 8 is the cube of 2. It is also one less than 9, making it the only cube that is one less than a square.

Linguistically, we have the curious phrase 'one over the eight', meaning drunk, with its implication that any reasonable person can down eight pints of beer without ill effects, but no more. Whether from Greek or Latin, eightsomes are indicated by the prefix oct- (as in 'octopus', of which the plural is 'octopuses' or 'octopodes', but definitely not 'octopi'). Strictly, words derived from the Greek ought to begin octa- with Latin ones starting octo-, but evidence of this has almost totally vanished. Here is an octad of more unusual oct- words listed in the *Oxford English Dictionary*: octaeterid (an eight-year period in the ancient Greek calendar), octarchy (government by eight rulers), octodactyl (eight-fingered), octoceratous (eight-horned), octodentate (eight-toothed), octogamy (marrying eight wives), octoglot (written in eight languages), octophthalmous (eight-eyed). The 'high octane' petrol you buy also refers to the eight carbon atoms in paraffin of the octacarbon series C_8H_{18}.

Eight is also the number of:

• bits in a byte in computer-speak
• bottles in a methuselah
• days in an ancient Roman week – seven working, one market
• English kings named Edward
• English kings named Henry
• feet in length of the largest normal koala's appendix
• furlongs in a mile
• kilograms of chocolate eaten per person per year in UK
• legs on a spider
• maids a-milking my true love gave to me
• ounces weight of a professional boxing glove
• the most mutton legs found in the stomach of one shark
• pints in a gallon
• qus (also spelt qs, ques or cues) in an old English penny – a vital piece of information for no-holds-barred Scrabble players, although 'qu' and 'qs' are both excluded from the official Scrabble list of permitted two-letter words
• reals in a Spanish dollar or peso; the coin was marked with the figure '8', hence the term 'pieces of eight'
• reindeer of Santa Claus: Dasher, Dancer, Prancer, Vixen, Comet, Cupid, Donner and Blitzen, first appearing in a poem 'A Visit from St Nicholas' written by Clement Moore in 1822. The more famous Rudolph was not created until 1939 in a story by Robert L. May
• years' marriage for a bronze anniversary

Here are eight films with an eight in their titles:

• *Dinner at Eight* (1933): John Barrymore, Lionel Barrymore and Jean Harlow drama
• *Curtain at Eight* (1934): murder in the theatre

• *Eight O'Clock Walk* (1954): Richard Attenborough as an English taxi driver wrongly accused of murder
• *Eight Witnesses* (1954): Dennis Price in a mystery where all the witnesses to a murder are blind
• *Where Eight Bells Toll* (1971): Jack Hawkins and Anthony Hopkins in an Alistair MacLean thriller
• *Kung Fu of the Eight Drunkards* (1980): martial arts
• *8 Men Out* (1988): classic tale of baseball corruption
• *8 Seconds* (1994): sports biopic

And let us not forget:
• *8½* (1963): Fellini's classic of love and art

9

'Nine worthies were they called, of different rites –
Three Jews, three pagans, and three Christian knights.'
Dryden, 'The Flower and the Leaf'

Dryden's Nine Worthies were Joshua, David and Judas Maccabeus; Hector, Alexander and Julius Caesar; King Arthur, Charlemagne and Godfrey of Bouillon. Nine is a popular number for such a list, for if good things come in threes what can be better than three sets of three? We meet another such nonad in Macaulay's 'Lays of Ancient Rome':

Lars Porsena of Clusium
By the nine gods he swore
That the great house of Tarquin
Should suffer wrong no more.

In this case, the reference was to the nine gods of the Etruscans: Juno, Minerva and Tinia (the three main gods), together with Vulcan, Mars and Saturn, Hercules, Summanus and Vedius. The nine lives of a cat, nine tailors it takes to make a man, the stitch in time that saves nine and the happiness of being on cloud nine are further examples of the goodness of nineness. The number does, however, have a darker side, with the ninth psalm foretelling the anti-Christ, and Christ himself dying on the cross at the ninth hour of the day (3 p.m. equals nine hours after dawn). The word 'noon' originally meant this ninth hour, but was later found to be more useful to designate the time we now know as midday.

Nine also carries connotations of worthiness in the phrase 'dressed up to the nines', though nobody is sure of its origin. One theory is that it has something to do with the 99th Foot Regiment of the British Army, which had a reputation for being better turned out than most; another idea links it to the Old English word eynes, meaning eyes. 'To then eynes' meant 'to the eyes'.

Mathematically, a number is divisible by nine if and only if the sum of its digits is divisible by nine. Nine is also three squared, and is the only square that is the sum of two consecutive cubes of positive integers: $9 = 3^2 = 1^3+2^3$.

Nine per cent of girls aged 16–18 say they would turn to their fathers for advice about sex; nine per cent of men do the ironing; and nine per cent of Britons drink neither tea nor coffee.

Nine is also the number of:
• babies born to eleven-year-old mothers in England and Wales in 1980
• circles of hell in the Divine Comedy
• circumference in feet of the cheese made for Queen Victoria in 1841 from a single milking of the 737 cows in Pennard, Glastonbury
• gallons of saliva secreted daily by a horse
• heads of the hydra, usually, though since Hercules found that two more grew for each one he cut off, the number was variable
• hours it takes an army ant to walk a mile
• ladies dancing that my true love gave to me
• milligrams of rat droppings permitted in a kilo of wheat by the US Food and Drug Administration
• Muses: Clio (history), Melpomene (tragedy), Thalia (comedy),

Calliope (epic poetry), Urania (astronomy), Euterpe (lyric poetry), Terpsichore (dance and choral song), Polyhymnia (song and oratory), Erato (love poetry)

• planets in the solar system: in order from the Sun, they are Mercury, Venus, Earth, Mars, Jupiter, Saturn, Uranus, Neptune, Pluto. If you need to remember this, just learn the following mnemonic: Men Very Easily Make Jugs Serve Useful Nocturnal Purposes

• points of the law, which according to Brewer comprise: a good deal of money; a good deal of patience; a good cause; a good lawyer; a good counsel; good witnesses; a good jury; a good judge; and good luck

• recorded accidents in the home in the UK in 1994 involving sash cords and in 1997 involving magnets

• width in inches of a cricket wicket

• words in the Solomon Islands for various stages of maturation of coconut

• years' marriage for a copper anniversary

There are only about 30 films with 'nine' in their titles, including:

• *Nine Days a Queen* (1934): Sybil Thorndike and Cedric Hardwicke in the story of Lady Jane Grey

• *The Man With Nine Lives* (1940): Boris Karloff as a mad doctor

• *Nine Lives are Not Enough* (1941): Ronald Reagan solves a murder

• *Nine Hours to Rama* (1963): film of the book by Stanley Wolpert

• *Nine Days of One Year* (1964): Russian drama about love, physics and radiation

• *9 to 5* (1980): Dolly Parton, Jane Fonda and Lily Tomlin plot revenge against male chauvinist boss

• *9½ Weeks* (1986): Kim Basinger in a tale of erotic obsession

• *9½ Ninjas* (1990): martial arts sex comedy

10

'Could I come near your beauty with my nails, I'd set my ten commandments in your face.'
William Shakespeare, *2 Henry VI*, 1592

The Pythagoreans had little doubt that ten was the most holy of all numbers. It was the sum of 1, 2, 3 and 4 and could thus be depicted as an equilateral triangle, known as the *tetraktys*, with one point at the apex, two arranged symmetrically below it, three below that and four along the base. With one representing existence, two creation, three life and four the elements from which everything is composed, ten just about summed up everything. All of which was, for the Pythagoreans, sufficient justification to explain the perfection of our having ten fingers and using that number as a base for our counting system. ('Decadactylous', incidentally, is the word for 'having ten fingers'.) In recent times, the *tetraktys* has been more commonly associated with tenpin bowling.

Linguistically, any of the prefixes deca-, decem- (both Latin) or deka- (Greek) signify groups of ten. Most of those listed in the *Oxford English Dictionary* relate to animals or plants, for example: decapterygious (ten-finned), decemcostate (ten-ribbed), decemdentate (ten teeth), decempedal (having ten feet, though at one time also used to mean ten feet long), decapetalous (ten petals) and decaspermal (producing ten seeds). For a fine example of the richness of the language, however, one could hardly do better than to have a choice between decaphyllous and decemfoliate, both of which mean having ten leaves. On a more human level, a ruling council of ten people is a decarchy.

Ten is also the number of:
- acres in a square furlong
- average duration in seconds of chimpanzees' intercourse
- British prime ministers who have served under Queen Elizabeth II

• chromosomes of a greenfly
• Commandments
• degrees Celsius of the average April temperature in London
• enquiries about John Dillinger's penis received annually by the Smithsonian Institution (it was reputedly thirteen inches long)
• events in a decathlon: 100m, 400m, 1500m, 110m hurdles, long jump, high jump, pole vault, shot-put, discus, javelin
• feet high a basketball ring should be
• grams in our average daily salt intake
• heliports in the UK
• lords a-leaping my true love gave to me
• passengers on Boeing's first commercial passenger plane, the 247
• place names in the US that include the word 'Breast'
• plagues of Egypt (Exodus 7–12): water becomes blood, frogs, lice, flies, cattle murrain, boils, hail and fire, locusts, darkness, slaying of first-born
• platforms at Moorgate and Baker Street stations – the most on the London Underground
• points on Mohs' scale of hardness: talc, gypsum, calcite, fluorite, apatite, feldspar (or orthoclase), quartz, topaz, corundum, diamond
• vowels in the Korean alphabet
• years' marriage for a tin anniversary
• years of age for the age of consent in England in 1576

Among the films that include 'ten' in their titles are:
• *The Ten Commandments* (1923): Cecil B. de Mille directs a cast of thousands in this silent epic
• *Ten Days That Shook the World* (1927): Eisenstein's epic silent film of the Russian Revolution
• *Ten Nights in a Bar-Room* (1931): the perils of alcohol
• *Kate Plus Ten* (1938): crime story from the book by Edgar Wallace
• *The Ten Commandments* (1956): Charlton Heston as Moses
• *Ten North Frederick* (1958): based on the novel by John O'Hara

• *Garlic is as Good as Ten Mothers* (1960): educational video
• *Under Ten Flags* (1960): Van Heflin and Charles Laughton wartime drama
• *10th Victim* (1965): cult sci-fi classic with Ursula Andress and Marcello Mastroianni
• *A Boy Ten Feet Tall* (1965): children's movie with Harry H. Corbett and Edward G. Robinson
• *Ten Little Indians* (1965): based on Agatha Christie's play *Ten Little Niggers*
• *10 Rillington Place* (1971): life of the murderer John Christie
• *Ten Days Wonder* (1972): Orson Welles in an Ellery Queen mystery
• *Hitler: The Last Ten Days* (1973): with Alec Guinness as Hitler
• *Ten Little Indians* (1975): the remake with Herbert Lom, Richard Attenborough and Charles Aznavour
• *10* (1979): starring Dudley Moore, Bo Derek and Ravel's Bolero

11

'At eleven o'clock he has his "elevenses", consisting of coffee, cream, more bread and more butter.'
P. G. Wodehouse, *Meet Mr Mulliner*

The derivation of the word 'eleven' is unknown, though some suggest it may come from the Aryan root *leiq* or *leip* meaning 'leave', on the tenuous grounds that eleven leaves you with one over, once you have counted up to ten. Another fact that has never been properly explained about eleven is why it is the number of players in a team for both Britain's major sports, cricket and soccer. One highly dubious explanation is that eleven, being one more than the holy number ten, is a reminder of the imperfection of mankind, especially when playing games. 'Eleven is sin; eleven transgresses the Ten Commandments,' wrote Schiller.

A more righteous side of eleven is reflected in the Dionysiads, a group of eleven women in Ancient Sparta formed to battle against the excesses of the Dionysian cult, and in the undecimvir, a panel of eleven Magistrates that sat in Ancient Athens. Undecim- is the prefix for anything to do with eleven, as in the words undeciman (or undecimarian) referring to church services that began at eleven o'clock, and undecimarticulate, having eleven sections or segments.

A modern though little realised example of undecimal counting is seen in the ISBN of published books. (It is, of course, a solecism to say 'ISBN number', since the N stands for 'Number' itself. The other letters are International Standard Book.) Any ISBN comprises ten digits. If you multiply the first by ten, the second by nine, the third by eight, and so on, summing the results as you go along, the result will always be divisible by eleven. This is a neat trick to guard against entering the wrong numbers on a computer. Any computer programmed to deal with ISBNs will perform the check automatically and stand a good chance of detecting any mistake. If you want to try this out on some of your own books, it may save time to remember the little test for divisibility by eleven: add the first, third, fifth and other odd digits of the number being tested and subtract the sum of the digits in the even places. If the result is divisible by eleven, then so is the number you started with.

Eleven is also the number of:

• airports in Albania, of which only five have paved runways

• the Apollo 11 spacecraft from which Neil Armstrong walked on to the moon

• cups of coffee drunk daily per capita in Sweden – the world's most prolific coffee-drinkers

• days lost in Britain in 1752 (3–13 September) on changing from the Julian to the Gregorian calendar (→365)

• golden hamster babies in an average litter

• hurricanes in the world in 1995 – the most since records began

• Oscars won by *Ben Hur* – the most for any film, a record equalled by *Titanic* in 1998. Curiously enough, eleven is also the record number of

Oscar nominations for a film that failed to win any, a lack of distinction shared by *Turning Point* (1977) and *The Color Purple* (1986)
• ounces in weight the average person loses overnight
• pipers piping my true love gave to me
• players in a soccer, hockey or cricket team
• recorded accidents in the home in the UK in 1994 involving drinking straws and in 1997 involving talcum powder
• riots at horse races in the US between 1960 and 1972
• states in the Confederacy
• years' marriage for a steel anniversary

The films with 'eleven' in their titles include:
• *Eleven Men and a Girl* (1930): US football romance
• *Ocean's Eleven* (1960): Shirley MacLaine, Frank Sinatra and Dean Martin comedy caper
• *11 Harrowhouse* (1974): comedy-crime caper with Trevor Howard, James Mason, Sir John Gielgud and Candice Bergen

12

'At twelve noon, the natives swoon, and no further work is done.'
Noel Coward, *Mad Dogs and Englishmen*

From the twelve months of the year to the twelve signs of the zodiac; from the twelve tribes of Israel to the twelve apostles of Christ, the number twelve has a powerful influence in a wide range of myths and cultures. Divisible by two, three, four and six, twelve would surely form a better basis for a counting system, if we did not fall two fingers short. Yet the usefulness of twelve is clearly shown in the existence of the words 'dozen' and 'gross'.

The number twelve has a curious arithmetical property: the square of 12 is 144 which, oddly enough, is the reverse of the answer when you square the reverse of 12: 21 squared equals 441. The only other pair of numbers with a similar property are 13 and 31.

Linguistically, the prefix for a dozen of anything is dodeca- as in the twelve-faced dodecahedron and the twelve islands in the Dodecanese group in the Aegean. (Though the 'Twelve Apostles' island group in Lake Superior comprises about twenty islands.)

The word 'twelve' is too old for anyone to know where it came from, but the initial tw- must have something to do with 'two'. Compared with 'eleven' the evidence is stronger that 'twelve' indicates 'two left' (when you have taken ten away). Brewer conjectures that at one time people had no need to count higher than twelve, so the numbers from one to twelve had distinctive names, with the -teen series introduced only when we had reached a 'more advanced state'.

Throughout the history of counting there have occasionally been suggestions that, in the spirit of easier divisibility, we ought to abandon the custom of counting to base ten and substitute numbers to base twelve. (So we would write twelve as 10, the number we now know as 24 would be 20 and the current 144 would be 100 in the new system.) In 1944, the Duodecimal Society of America was founded in order to campaign for such a change, and the Duodecimal Society of Great Britain followed in 1959. Both subsequently changed the 'Duodecimal' in their names to the simpler 'Dozenal'. Despite general agreement on their objectives, the two Dozenal societies have had long-running differences about how to write the new numbers necessary to signify the old ten and eleven, and what to call them. The original proposal of the DSA was to have X for ten and E for eleven, but the British preferred to use 2 and 3 written upside down, which they borrowed from Sir Isaac Pitman's *Phonetic Journal*. Then the Americans changed to X and #. There have also been differences of opinion on the words to be used for the new numbers and whether the new '10' should be called 'ten' or 'one doz'. The Dozenal Society of America have adopted as their mascot the panda, because a panda is widely supposed to have six fingers on each paw. In fact it has only five fingers, but uses a protuberance on the side of its wrist rather like an extra digit.

Twelve is also the number of:
• affairs Earl Spencer was accused of having during his divorce case
• apostles, chosen by Jesus to spread his teachings after his death. Originally they were: Andrew, Bartholomew (or Nathaniel), James son of Alphaeus, James son of Zebedee, John, Jude (or Thaddeus), Judas Iscariot, Matthew (or Levi), Philip, Simon Peter, Simon the Zealot, and Thomas. After Judas's suicide, he was replaced by Matthias. St Paul is also generally included, because of his claim to have seen Jesus after the resurrection
• astrological houses: Life, Fortune & Riches, Brethren, Parents & Relatives, Children, Health, Marriage, Death, Religion, Dignities, Friends & Benefactors, Enemies
• bottles in a salmanazar
• Days of Christmas
• drummers drumming my true love gave to me
• 'filthy words' laid down by Federal Communications Commission in 1973: fuck, shit, piss, cunt, turd, twat, fart, ass, motherfucker, cocksucker, tits, cock
• gallons of saliva secreted daily by the average cow
• grams of instant coffee consumed by the average Briton each week
• inches in a foot
• hours each month the average American spends in shopping malls
• Labours of Hercules, performed to expiate his guilt after killing his wife and children in a fit of madness. They were: killing the Nemean lion and the Hydra of Lerna, capturing the Hind of Ceryneia and the Boar of Erymanthus, cleaning the Augean stables, chasing away the Stymphalian birds, capturing the Cretan bull and the horses of Diomedes, stealing the girdle of Hippolyta, capturing the oxen of Geryon, stealing the apples of the Hesperides, and capturing and binding Cerberus in Hades
• letters in the Hawaiian alphabet

• metres height of the average Christmas tree in a public square in the US
• minutes a bedbug takes to feed
• pairs of ribs of a human being
• pence in an old shilling
• people killed on average each year in Hong Kong by rubbish thrown from tall buildings
• people who have set foot on the moon
• percentage of the Swiss population who are officially obese
• points on the Mercalli Scale of Felt Intensity of earthquakes (from 1 – perceived only by sensitive instruments – to 12 – masses of rock displaced horizontally and vertically)
• points on the original Beaufort scale of wind intensity
• postal deliveries a day in London at the end of the nineteenth century
• randomly tuned radios in John Cage's *Imaginary Landscape No. 4* (1953)
• riots at motor sport events in US between 1960 and 1972
• signs of the zodiac
• square kilometres of Africa always covered in ice
• tablets on which the 'Laws of the Twelve Tables' were inscribed. The first written laws of the Romans, they were fixed to the speaker's stand in the Forum
• tons of jellybeans bought by the White House during the Reagan Presidency
• Tribes of Israel
• tricks for a small slam in bridge
• years' marriage for a silk (or fine linen) anniversary

Apart from *The Dirty Dozen* (1967), and *The Dirty Dozen: The Next Mission* (1985), and *The Dirty Dozen: the Deadly Mission* (1987), and *The Dirty Dozen: The Fatal Mission* (1988), the films that come in twelves include:

• *Twelve Miles Out* (1927): silent film of drama aboard a ship with Joan Crawford

- *Twelve O'Clock High* (1942): war film with Gregory Peck
- *Her Twelve Men* (1954): based on a book *Miss Baker's Dozen* by Louise Baker about a teacher with thirteen mischievous boy pupils
- *Twelve Angry Men* (1957): juryroom drama with Henry Fonda and Lee J. Cobb
- *The Twelve-Handed Men of Mars* (1964): Italian sci-fi
- *The Twelve Chairs* (1970): Mel Brooks's version of the Ilf and Petrov story

And look out for:
- Play: *The Twelve-Pound Look* (1910) by J. M. Barrie
- Book: *The Wind's Twelve Quarters* (1975) by Ursula Le Guin

13

'It was a bright cold day in April, and the clocks were striking thirteen.'
George Orwell, *1984*

Thirteen is widely considered to be an unlucky number, though the origins of that fear – correctly termed triskaidekaphobia – are unclear. In Christian tradition, fear of thirteen is usually linked to the number present at the Last Supper, though that superstition dates back only to the Middle Ages by which time thirteen-fear was already widely established in other cultures. The Babylonians were suspicious of thirteen, apparently through a feeling that the sun needed to be kept separate from the twelve zodiac signs. Old Norse mythology tells of a banquet in Valhalla at which Loki arrived as an uninvited guest, making the number up to thirteen, which resulted in the death of Baldur.

The 1894 edition of Brewer's *Phrase and Fable* tells us that 'the Turks so disliked the number that the word is almost expunged from their vocabulary' and 'the Italians never use it in making up their lotteries'. The wisdom of the Italians was confirmed more than a century later when, after the first year of the British National Lottery, the two numbers that had been least favoured by the draw turned out to be 13 and 39 (3×13) – though 26 (2×13) had done considerably better than average. Around the end of the nineteenth century, street numbers in Paris would always jump from 12 to 14, omitting the unlucky number, while blocks of flats in London tended to have the euphemistic 12A. Even as late as 1960 the newly-built Carlton Towers hotel in London's Knightsbridge considered it wise to jump directly from the 12th to the 14th floor and in 1998, when the Citibank tower was being built in London's Docklands, the thirteenth storey of the structure was clearly identified as '12+1' during the building.

Writing in the *Smithsonian Magazine* in 1987, Paul Hoffman estimated that fear of the number thirteen costs the United States 'a billion dollars a year in absenteeism, train and plane cancellations and reduced commerce on the thirteenth of the month'. In 1967, a group of thirteen Americans were reported to have launched a campaign to rid the country of triskaidekaphobia. Their first act was to rent a plot of land thirteen feet long for thirteen cents a month. One noted triskaidekaphobe was the composer Arnold Schoenberg (pioneer, incidentally, of twelve-tone music), who died, as he had predicted, at the age of 76 (7+6 = 13), supposedly on Friday the thirteenth (→688) at thirteen minutes to midnight.

French society used to support a group of noble gentlemen known as the *quatorziennes*, who made themselves available at short notice to attend any dinner party or other formal function at which exactly thirteen people had turned up. Others, however, have – for good reasons of their own – associated thirteen with good luck. King Louis XIII of France was so fond of the number that he married Anna of Austria when she was thirteen years old, while world chess champion Garry Kasparov was born on the thirteenth of the month, scored his first ever victory against his great rival Anatoly Karpov on the thirteenth of the month, and has scored particularly well in thirteenth games of world championship matches.

Thirteen is also the number of:

- cards in a suit
- Coca-Colas a day sold in first eight months of its existence at Jacob's pharmacy in Atlanta, Georgia
- estimated percentage of Britons trying to slim
- inches waist measurement decreed for ladies at court of Catherine de Médici
- lines in a rondeau
- original colonies of the US
- pounds' weight loss of an average mother during childbirth
- record number of Caesarian births by one woman
- recorded accidents in the home in the UK in 1994 involving rubber bands or elastic and in 1997 involving draught excluders
- stripes on the American flag (one for each of the original colonies)
- times the spoil from digging the Channel tunnel would fill Wembley stadium
- tricks for a grand slam in bridge
- years of age at which the average American develops a phobia
- years' marriage for a lace anniversary

The ill-omened connotations of thirteen have made it a popular number in films, including the highly successful horror film, *Friday the Thirteenth* (1980) and its seven sequels. Other less horrific films include:

- *13 Washington Square* (1928): silent crime/romance comedy
- *13 Rue Madeleine* (1946): D-Day spy drama with James Cagney and E. G. Marshall
- *13 Lead Soldiers* (1948): Bulldog Drummond story
- *The 13th Letter* (1951): Michael Rennie is accused of having an affair with Charles Boyer's wife
- *13 West Street* (1962): Alan Ladd and Rod Steiger in action drama based on the book *The Tiger Among Us* by Leigh Brackett
- *The 13 Chairs* (1970): Terry-Thomas and Orson Welles in Ilf and

Petrov Russian folk tale (also filmed by Mel Brooks as *The Twelve Chairs*)
• *In a Year of Thirteen Moons* (1980): sex-change drama by Rainer Werner Fassbinder
• *Thirteen at Dinner* (1985): murder mystery with Faye Dunaway and Peter Ustinov
• *The Alamo: Thirteen Days to Glory* (1987): Davy Crockett's last stand
• *The 13th Mission* (1991) US soldiers fight their way out of an Asian jungle

14

'I spent fourteen months at Magdalen College: they proved the fourteen months the most idle and unprofitable of my whole life.'
Edward Gibbon, *Memoirs of My Life*, 1796

Despite the decline and fall of Gibbon's time at Magdalen College, the number fourteen is generally considered propitious. The moon takes fourteen days to change from new to full, whence the significance of a fortnight. The words of a popular old German lullaby run: 'When at night I go to sleep, fourteen angels watch do keep,' which indicates the protective quality of fourteen. It was the ancient Egyptians, however, who seemed most interested in fourteen. The great god Osiris, judge of the dead and potentate of the kingdom of ghosts, was identified with the moon, so it should be no surprise to hear that he is represented by an eye at the top of fourteen steps. When he was killed by his brother, Set, Osiris was cut into fourteen pieces which were buried in different places, each bringing the blessing of fertility to a different region. The myth goes that Isis, who was both Osiris's sister and his wife, sought out the pieces in order to reunite them, and found all but one. The fourteenth piece, the mystical 'Talisman of Set', was Osiris's penis.

For the best mysticism involving the number fourteen, however, we must go the French kings. Louis XIV ascended the throne in 1643

(1+6+4+3 = 14), reigned for 77 years (7+7 = 14), died in 1715 (1+7+1+5 = 14) and, if we add his birth year (1638) to his death year, the answer is 3,353, again adding up to fourteen. Henri IV, on the other hand, was not an XIV himself, but he was the fourteenth king of France and his name, Henri de Bourbon, had fourteen letters. He was born on 14 December 1553 (1+5+5+3 = 14) and was assassinated on 14 May 1610 at the age of fifty-six, which is four times fourteen.

The correct term for a set of fourteen things is a tessaradecad, not to be confused with a tetradecapod, which is a creature with fourteen feet. A poem of fourteen lines we all know to be a sonnet, but only as long as it fulfils certain other strict conditions, such as one of a set of prescribed rhyming schemes and having ten syllables in every line. A fourteen-line poem that does not quite qualify as a true sonnet may be known as a decatessarad or a quatorzain. A line containing fourteen syllables is known as a tessaradecasyllabon or, more simply, a fourteener.

There are fourteen pounds in a stone, which seems curious, since fourteen has neither the merit of being a round decimal number nor the convenience of allowing easy subdivision in a way that the twelve pence in a shilling or sixteen ounces in a pound would allow. The explanation lies in the ramshackle origins of English pre-metric weights and measures. A pound (derived from the ancient Roman *libra pondo*, pound in weight, from the first word of which the lb abbreviation originates) used to vary between twelve and twenty-seven ounces, depending on the commodity that was being weighed. The sixteen-ounce pound *aveir de peis* (of merchandise weight) was made standard for bulky commodities by Edward III. Since 1826 this has been the only legal pound. Meanwhile, however, a stone varied between eight and twenty-four pounds. The fourteen-pound stone was decreed as standard by a 1495 Act of Henry VII, but only when wool was being weighed. Even as late as 1835 an Act of William IV declared: 'By local Customs, the Denomination of the Stone Weight varies'. The whole system only finally attained some degree of order with the Weights and Measures Act of 1878.

Fourteen is also the number of:

• countries bordering Russia
• days an ant can survive underwater
• days' incubation for measles
• days' reign of Lady Jane Grey: from coronation to execution, she was Queen of England for a period of only nine days, from 10 to 19 July 1553, but she had accepted the crown reluctantly on the death of Edward VI on 6 July. Her reign could therefore be said to extend from 6 to 19 July, which is 14 days in all
• impulse decisions made by average US supermarket shopper
• inches in the length of an average okapi's tongue
• kisses in the novels of Jane Austen (of which four are between females, four involve men kissing women on the hand, two bestowed by women on children, one given by a man to a lock of severed hair and three involving lip contact between man and woman)
• maximum number of golf clubs permitted per player
• minutes each day the average American spends reading magazines
• mph speed limit for light locomotives under the Highways Act of 1896
• news weeklies on sale in London in 1645
• percentage of Londoners who have talked to their neighbours in the past week
• percentage of the world's sheep living in Australia and the world's onions grown in China
• possible meanings of the expression 'make up', according to a 1933 book entitled *American Speech*
• seconds it took Napoleon's surgeon, Baron Dominique Larrey, to amputate a leg
• times Nuri as-Said served as prime minister of Iraq between 1930 and 1958
• vernacular and independent tongues of Europe, according to a seventeenth-century linguist named Howell
• wattage of a human brain in deep thought

- weight in ounces of the largest hen's egg on record
- years' marriage for an ivory anniversary
- years of age at which Rin-Tin-Tin died in 1932

Films and other works with fourteen in their titles include:

- *Fourteen Hours* (1951): from a story by Joel Sayre called *The Man on the Ledge*. Grace Kelly made her screen debut in this film
- *The Fourteen* (1973): youthful drama with Alun Armstrong
- *Flying Claw Fights Fourteen Dragons* (1980): adventure
- *14 Going on 30* (1988): teenage romance comedy with Loretta Swit
- Poetry: 'Fourteen Men' (1954) by Dame Mary Gilmore

15

'And fifteen arms went round her waist
(And then men ask "Are barmaids chaste?")'
John Masefield, *The Everlasting Mercy*, 1911

The Germans used to have a measure called a 'Mandel', referring to a collection of fifteen small objects. The word means 'little moon' and refers to the fifteenth day in a lunar month when the moon reaches its full strength. In 1851 Sir Edward S. Creasy wrote *The Fifteen Decisive Battles of the World* in which he listed the following as the most significant military battles in history:

1. The battle of Marathon (490 BC) when Miltiades, commanding 10,000 Greeks, defeated a Persian army of 100,000.
2. The battle of Syracuse (413 BC) when the Athenians were defeated with the loss of their entire fleet and 40,000 killed or wounded.
3. The battle of Arbela (331 BC) when Alexander the Great defeated Darius Codomanus.
4. The battle of Metaurus (207 BC) when Livius and Nero destroyed

Hasdrubal's army, which had been sent to reinforce Hannibal.

5. The defeat of the Romans by Arminius and the Gauls in AD 9, thus establishing the independence of Gaul.
6. The battle of Chalons (AD 451), at which Attila was defeated.
7. The battle of Tours (AD 732), when the Saracens were defeated and Europe broke free of the Moslem influence.
8. The battle of Hastings (1066) when William the Conqueror defeated Harold II.
9. The battle of Orléans (1429) when Joan of Arc secured independence for France.
10. The defeat of the Spanish Armada in 1588.
11. The battle of Blenheim (1704), when Marlborough and Prince Eugene inflicted the first major defeat on the forces of Louis XIV.
12. The battle of Pultowa (1709) when Czar Peter routed the forces of Charles XII of Sweden, thus laying the basis for Russian power.
13. The battle of Saratoga (1777) in the American War of Independence, when a large British army under John Burgoyne was defeated by American troops under the command of Horatio Gates.
14. The battle of Valmy (1792) when Marshal Kellerman of France defeated the Duke of Brunswick.
15. The battle of Waterloo (1815) when Napoleon was defeated by Wellington.

Fifteen is also the number of:

• babies in the biggest recorded litter by a ferret
• blows by the executioner to sever the head of Mary Queen of Scots
• escalators at Baker Street tube – the most on any station of the London Underground
• inches diameter of netball ring
• inches height of pins in tenpin bowling
• letter Os in an Italian Scrabble set
• medals awarded to Audie Murphy, most decorated US soldier in

World War II
• full member states of the European Union
• men on a dead man's chest sung of in Robert Louis Stevenson's *Treasure Island*. He borrowed the 'Dead Man's Chest', incidentally, from Charles Kingsley, who used the phrase as the name of one of the Virgin Islands in his novel *At Last* which was published in 1871, ten years before *Treasure Island*
• minimum number of checkouts in a hypermarket
• minutes everyone will be famous, according to Andy Warhol
• people per square kilometre in Finland
• pieces on each side in backgammon
• pounds of hydrogen in the average human body
• red balls in snooker
• sets of beads (one large and eleven small in each set) on a rosary, traditionally connected with fifteen mysteries of the life of Mary
• years of age of Sheridan's bashful maiden: 'Here's to the maiden of bashful fifteen' (*The School for Scandal*)
• years of the Fifteen Years' War between Austria and the Ottoman Turks from 1591 to 1606
• years' marriage for a crystal anniversary

The films and other works with 'fifteen' in their titles include:
• *15 Maiden Lane* (1936): crime feature with Cesar Romero
• *Angels One Five* (1954): war drama
• Poetry: 'Fifteen Dead' (1979), by Thomas Kinsella

And let us not forget that every 15 seconds, someone somewhere in Britain starts to dig a hole in the road

16

'Anni? Sedecim. Flos ipsus.'
Terence (Publius Terentius Afer) *c.*160 BC
(Age? Sixteen. The very flower of youth.)

Sixteen is the fourth power of two, which is why computers find it so useful as a base for counting. Indeed, much computer code is written in hexadecimal notation, with the digits from zero to nine augmented by the letters A to F. In the nineteenth century, J. W. Mystrom suggested that life would be easier if we all counted using base sixteen. Instead of one, two, three and so on, he proposed renaming the numbers an, de, ti, go, su, by, ra, me, ni, hu, vy, la, po, fi, ton. The idea, however, never caught on.

The delights of sixteen were appreciated by the ancient Romans, whose measuring system had four fingers to a hand and four hands to a foot. Thus there were sixteen fingers to the foot. The ancient Indians also thought well of the number, listing sixteen parts of a complete human, sixteen signs of beauty and sixteen pieces of jewellery adorning a perfectly bejewelled lady.

The Pythagoreans were fond of sixteen, as it is the only number that measures the perimeter and area of the same square.

The number also has a role to play in some European languages. Just as the counting systems in English and German change gear after twelve and start systematically adding -teen or -zehn to create higher numbers, the French and Italians persist with original number names until 16 (*seize, sedici*), before adopting the more consistently formulaic (*dix-sept, diciasette*) style for the remaining teens.

Sixteen is also the number of:

• the amendment to the US constitution that introduced Federal Income Tax in 1913

• annas in a rupee in India before they changed to decimal currency in

1957. The term derived from a Hindi word *ana* meaning insignificant. Despite its insignificance, an anna could be divided into four *pice*
• arms of the goddess Pussa
• bottles in a balthazar
• centimetres length of the average erect penis
• chromosomes of the honey bee
• countries bordering China
• days in the gestation period of a golden hamster
• drams in an ounce, ounces in a pound
• eggs in the egg-purse of a cockroach
• feathers in the tail of a red or spotted grouse
• language groups of the 133 Indian tribes recognised in Mexico in 1914
• mingles in a stekamen (Dutch liquid measure of the eighteenth century)
• murders in Atlanta during the Olympic Games in 1996
• orders of mammals
• pages the Pentagon's Dept of Food Procurement takes to define plastic whistles
• percentage of the world's silver from Mexico and the world's grapes grown in Italy
• pounds weight minimum for a shot used in shotputting
• rules in Esperanto
• stitches in an embroiderer's eye stitch – all entering a single hole, but spread into a square on the outside edge
• syllables in each line of a sloka – a couplet of Sanskrit verse
• violin concertos by Vivaldi that Bach arranged for harpsichord
• years you have to be married for there to be no metal or other substance conventionally associated with the anniversary – you must now wait four more years (→20) for a piece of china

Films and other works that celebrate the number sixteen include:

• *Sixteen Fathoms Deep* (1948): from a story by Eustace Adams called

Sixteen Fathoms Under
• *Sixteen* (1972): coming-of-age story of girl from the deep south
• *Sweet Sixteen* (1983): psychic horror with Bob Hoskins and Susan Strasberg
• Song: 'Sweet Little Sixteen' (1958) by Chuck Berry

And let us not forget Sixteen-string Jack, alias John Rann, an eighteenth-century highwayman. He wore sixteen tags, eight at each knee and his stylish, if foppish manner even aroused the admiration of Samuel Johnson: 'Dr Johnson said that Gray's poetry towered above the ordinary run of verse as Sixteen-string Jack above the ordinary foot-pad.' (Boswell, *Life of Johnson*.) His career came to a sad end when he was hanged in 1774.

17

'I kiss'd her slender hand,
She took the kiss sedately;
Maud is not seventeen
But she is tall and stately.'
Tennyson, *Maud*, 1855

If you are searching for a mystical significance of the number seventeen, you need look no further than the Great Flood in the Book of Genesis, which began on the seventeenth day of the second month and ended on the seventeenth day of the seventh month. The watery nature of seventeen is also reflected in Greek legend, for that is the number of days Odysseus floated on a raft after leaving Calypso. Perhaps because of this the Greeks considered seventeen to be an unlucky number. (See 153 for further symbolism connected with the number seventeen.)

If you cube the number 17 you get 4,913, which, if you add its digits together, gets you back to 17. The only other numbers that are equal to the sum of the digits of their cubes are 1, 8, 18, 26 and 27.

Seventeen is also the number of:
• bottle banks in the UK in 1977 (the number had risen to over 16,000 by 1997)
• children produced by Queen Anne
• consecutive draws played by Anatoly Karpov and Garry Kasparov in their world chess championship match in 1984
• feet width of a badminton singles court
• goldfish bowls left on public transport in Tokyo in 1979
• inches of a human bottom allowed in the specification for seat width for the British Rail 'Sprinter' train
• Italians expelled from Great Britain 1907–14 for prostitution and procuring
• laughs a person has each day on average
• the line of latitude dividing N. and S. Vietnam
• minutes a day the average American spends reading books
• minutes a day the average Japanese father spends looking after his children
• most bananas eaten in two minutes
• muscles controlling a dog's ear
• pages the Pentagon's Dept of Food Procurement takes to describe olives
• points in the Seventeen Article Constitution, Japan's first known written law, introduced by Prince Shotoku in 604
• points on the Beaufort Scale of wind strength since 1955. Sir Francis Beaufort (1774–1857) had only 12 points on his original scale. The US Weather Bureau added five more when accurate measurements of hurricane force winds became possible
• syllables in a haiku – which probably stems from an ancient Japanese belief that seventeen is the optimum number of syllables spoken in a single breath
• times James Boswell suffered from gonorrhoea

- tons of gold made into wedding rings each year in the US
- tributaries of the River Severn

Seventeen is a rare visitor to film titles, with these two among the exceptions:

- *Number Seventeen* (1932): early Hitchcock jewel-robbery drama
- *Seventeen* (1940): comedy based on the book *Seventeen* (1916) by Booth Tarkington

18

'I've had eighteen straight whiskies – I think that's the record.'
Allegedly the last words of Dylan Thomas, 1953

Eighteen was a mystic number for the Mevlevi – the Sufi order of whirling dervishes – whose apprenticeship included learning the eighteen different kinds of service in the kitchen. It is also the age at which adulthood traditionally begins and is thus when, in England and Wales at any rate, you can sign a cheque, buy a house, buy and drink alcohol, serve on a jury, have a credit card, get married without parental consent, make a will and be hanged.

We have already mentioned that eighteen is equal to the sum of the digits in its cube (5,832), but the number exhibits a unique property if we move on to its fourth power (104,976). For 18^3 and 18^4 together utilise each of the digits from 0 to 9 once each.

Linguistically, 'eighteen-wheeler' is CB-radio speak for a juggernaut.

Eighteen is also the number of:

- the Amendment to the US constitution introducing Prohibition
- feet in the diameter of the pitcher's circle in baseball
- French kings called Louis
- haircuts performed in sixty minutes by a Southampton hairdresser

Trevor Mitchell to set a new record on 28 October 1996
• holes on a golf course
• inches in a cubit, approximately, though the length varied as it was defined as the distance from an individual's elbow to the tip of his middle finger
• inches in one mkono (E. African unit of length)
• inches in the diameter of a basketball ring
• islands in the Faeroes
• known satellites of Saturn
• letters in the words 'conversationalists' and 'conservationalists', the longest pair of anagrams (excluding scientific terms) in English
• pairs of ribs of a horse
• people in the UK per personal computer bought in 1994
• players in an Australian Rules football side
• pounds in the record weight of a Brussels sprout
• Russians and Poles expelled from Great Britain between 1907 and 1914 for prostitution and procuring
• things known by Odin in old Scandinavian mythology
• times the word 'dog' appears in the Bible
• tons of smoked salmon eaten at Wimbledon in 1990
• weeks at number one for Frankie Laine's 'I Believe'

Eighteen is also the age at which one can go to see any film in the cinema, which may account for its popularity in film titles including:
• *Under Eighteen* (1932): teenage romance with Regis Toomey
• *Nearly Eighteen* (1943): comedy
• *Eighteen and Anxious* (1957): love and disappointment in Las Vegas
• *18 in the Sun* (1964): teenage romance
• *18 Again!* (1988): George Burns as an 81-year-old who exchanges souls with his 18-year-old grandson

And look out for:

• *Eighteen Poems* (1934), the first published book by Dylan Thomas – which sums up his career rather neatly: from eighteen poems to eighteen whiskies.

19

'At nineteen, you know, one does not think very seriously.'
Jane Austen, *Emma*, 1815

Nineteen years is the period of the Metonic cycle, named after the Greek astronomer Meton, who discovered it in 433 BC. After one full cycle, the phases of the moon recur on the same dates – in other words, it's the time you have to wait before your diary will again be correct in its listing of the full moons. The Metonic cycle was responsible for the entry into the language of the word decennoval (or decennovenal) which means of, or pertaining to, a period of nineteen years.

The number nineteen has great significance in the Quran, according to some interpretations. Indeed, one analysis uses items connected with the number 19 in a 'proof' that the Quran is truly the word of God. The starting point is a single verse that states 'Over it is nineteen' in a section relating to the number of angels guarding Hell. Many scholars over the centuries have argued that this means there are nineteen guardians of Hell, despite the fact that God in the very next verse told them that nobody other than He knows the number of the angels of Hell. Quranic word-counters, however, have unearthed enough nineteens to convince them that the number has a deep significance. Here are the first few items from a recent inventory of Quranic nineteens:

1 The first verse consists of 19 Arabic letters
2 Each of the four Arabic words in the first verse is repeated in the Quran in multiples of 19 in numbered verses:
The first word 'Ism' (Name) occurs 19 times

The second word 'Allah' (God) occurs 2698 times (which equals 19×142)
The third word 'Al-Rahman' (Most Gracious) occurs 57 times (19×3)
The fourth word 'Raheem' (Most Merciful) occurs 114 times (19×6)

3. The Quran consists of 114 suras, which is 19×6
4. The total number of verses in the Quran is 6346, or 19×334. Also 6+3+4+6 = 19

Nineteen is also a sacred number in the Baha'i faith, in which the year is divided into nineteen months of nineteen days each. Shakespeare, however, was distinctly less fond of the number nineteen, using it only three times in his entire work, fewer than any other small number.

Nineteen is also the number of:

- acres of pizza eaten daily in the US
- boxing riots in the US between 1960 and 1972
- cheeses that can be made from 9 gallons of goat's milk according to Aristotle
- days in the gestation period of a laboratory mouse
- grades of pencil from 9H to 9B with HB in the middle separating the Hs from the Bs
- grams in the average weight of a Chinese man's testicle
- guns in a salute for the Vice-President of the United States
- Hungarian rhapsodies by Liszt
- James Bond movies
- largest ordinal adverb in the *OED* – firstly, secondly, thirdly up to nineteenthly are listed, but not twentiethly
- letter As in a Malaysian Scrabble set, the most of any letter in any language
- percentage of the world's goats living in India
- percentage of the world's pineapples from Thailand
- stations on the Baku underground in Azerbaijan

Finally, we must mention the curious phrase 'nineteen to the dozen' to mean at a very fast rate, though why nineteen is chosen rather than, say, eighteen or twenty is a matter for speculation. One proposed explanation links the numbers to the efficiency of pumping engines in the coal or tin mining industry, with the best engines yielding 19 tons of coal per 12 gallons of fuel. It seems more likely, however, that the numbers were chosen simply as a euphonious example of an exaggerated rate. In Welsh, they say 'Siaradai'r hen wraig pymtheg yn y dwsin bob amser' – The old woman always talked fifteen to the dozen.

The only film we have spotted with '19' in its title is:

• *Montparnasse 19* (1957): biographical drama on the last weeks of Modigliani

Song:

• '19th Nervous Breakdown' was a hit for the Rolling Stones in 1966. It reached number two in both the UK and the US charts – which is nineteenth place counting up from the bottom of the top twenty

20

'Then come kiss me, sweet and twenty,
Youth's a stuff will not endure.'
William Shakespeare, *Twelfth Night*, 1601

Twenty, as two times ten, plays a significant role in our counting systems, as is seen in the French *quatre-vingts* (four-twenties) to mean eighty and the English 'score' to mean twenty of anything. The latter probably derives from the practice of counting herds of sheep or cattle in twenties, perhaps keeping tally by making a notch or 'score' on a piece of wood. The lifespan allotted to us in the Bible – 'three score years and ten' – and Abraham Lincoln's 'four score and seven years ago' confirm that counting in twenties used to be reasonably widespread. Around the fourth century AD, the Mayans used a counting system with base twenty (properly termed 'vigesimal'). Old German also used to have a word *Schneise* which meant a rope on which twenty codfish were left to dry.

Other words that indicate the number twenty include 'vicenary' (of a group of twenty), 'vicennial' (of a period of twenty years) and anything beginning with the prefix 'icosa-'. The regular icosahedron, for example, has twenty faces, each of which is an equilateral triangle, and icosasenic means lasting the same time as twenty short syllables. Icosian simply means related to twenty.

And talking of being related, twenty is the number of specific relations a Briton is forbidden to wed. From 1662 until the Marriage Act of 1949, the law was based on the Table of Affinity and Kinship in the Book of Common Prayer which listed thirty forbidden relationships: A Man could not marry his Grandmother, Grandfather's Wife, Wife's Grandmother, Father's Sister, Mother's Sister, Father's Brother's Wife, Mother's Brother's Wife, Wife's Father's Sister, Wife's Mother's Sister, Mother, Step-Mother, Wife's Mother, Daughter, Wife's Daughter, Son's Wife, Sister, Wife's Sister, Brother's Wife, Son's Daughter, Daughter's Daughter, Son's Son's Wife, Daughter's Son's Wife, Wife's Son's Daughter, Wife's Daughter's Daughter, Brother's Daughter, Sister's Daughter, Brother's Son's Wife, Sister's Son's Wife, Wife's Brother's Daughter, Wife's Sister's Daughter. The list of relatives a woman could not marry was totally analogous. The system was rationalised and simplified in 1949, cutting the number to twenty. Among other freedoms, a man may now marry his father's brother's wife or his wife's mother's sister.

Twenty is also the number of:
- bottles in a nebuchadnezzar
- feet from side to side in a badminton doubles court
- hot dogs eaten in twelve minutes by Steve 'The Terminator' Keener to win Nathan's annual hot dog eating contest in New York in 1999.

• 58 novels by Rider Haggard

This fell well short of the twenty-four and a half hot dogs world record, held by Hirofumi Nakajima
• hundredweight in a ton
• lascivious turtles mentioned in the *Merry Wives of Windsor* (*'Well, I will find you twenty lascivious turtles ere one chaste man.'*)
• major golf championships won by Jack Nicklaus
• numbers round the edge of a dart board
• pages the Pentagon's Dept of Food Procurement takes to describe hot chocolate
• points for a minor suit trick in contract bridge
• pounds in the record weight of a cucumber
• pounds of rabbit meat eaten per head of population each year in Malta
• questions in the famous quiz game
• square feet in area of an average human's skin
• tons weight of the average iceberg
• Wimbledon titles won by Billie-Jean King
• years' marriage for a china anniversary
• years Rip van Winkle slept in the story by Washington Irving

Films and other works celebrating the number twenty include:
• *20-Mule Team* (1940): adventures of borax miners in the Old West
• *Love at Twenty* (1963): romance directed by Andrzej Wajda and François Truffaut
• *20 Shades of Pink* (1976): worried artist drama with Eli Wallach and Anna Jackson
• *Twenty Bucks* (1993): comedy with Spalding Grey and Christopher Lloyd
• *Twenty Years After* (1845): Alexandre Dumas's sequel to *The Three Musketeers*
• *Twenty Glances at the Infant Jesus* (1944), a piano work by Olivier Messiaen

21

'When I was one-and-twenty
I heard a wise man say
"Give crowns and pounds and guineas
But not your heart away;
Give pearls away and rubies,
But keep your fancy free."
But I was one-and-twenty,
No use to talk to me.'
A. E. Houseman, *A Shropshire Lad*, 1896

Until 1970, in Britain, you had to wait until you were twenty-one to do most of the things you can now do when you are eighteen. An age of majority of twenty-one, however, still remains for certain aspects of life including becoming an MP, taking your seat in the House of Lords, and adopting a child.

Why twenty-one was chosen in the first place as the age of majority remains something of a mystery, though in the light of the mystic connections of the numbers three and seven perhaps their product might be expected to have attained a special significance.

Twenty-one is also the number of:
• gallons of beer drunk per capita each year in Australia
• guns in a salute for the American President. This practice originated as a royal salute in Britain from ships of the line, which had a maximum of twenty-one guns along one side of the vessel
• leap seconds added to time since 1971 to correct atomic clocks and keep them in time with the slowing of the earth's rotation
• letters in the language of the angels, according to Elizabethan mystics John Dee and Edward Kelly
• piano concerti by Mozart – although the numbering of such concerti goes up to 27, several of the early ones were no more than

transcriptions of works by other composers
• pips on a die
• republics of the Russian Federation
• shillings in a guinea
• years of the Twenty-One Years War, from 1701 to 1721, between Russia and Sweden

Films with twenty-one in their titles include:
• *Free, Blonde and Twenty-One* (1940): comedy romance
• *Twenty-One Days Together* (1940): three weeks in the company of Robert Newton and Laurence Olivier
• *21 Hours at Munich* (1976): TV film drama based on the Arab terrorist murders at the 1972 Olympics
• *21 Jump Street* (1987): Johnny Depp in a high-school crime drama
• *Twenty-One* (1991): comedy romance with Patsy Kensit and Patrick Ryecart

22

'For two-and-twenty sons I never wept,
Because they died in honour's lofty bed.'
William Shakespeare, *Titus Andronicus*, 1594

Twenty-two is a number with deep mystic significance because it has been calculated to be the number of things made by God in the six days of creation: on Day One: unformed matter, angels, light, the upper heavens, earth, water, air; Day Two: the firmament; Day Three: the seas, seeds, grass and trees; Day Four: the sun, moon and stars; Day Five: fish, aquatic reptiles and flying creatures; Day Six: wild beasts, domestic animals, land reptiles and man. That list was first given by Isidor of Seville in the seventh century, and bears all the ingenuity of a man determined to draw up a list with the same number of elements as there are letters in the Hebrew alphabet. So God made the twenty-two things comprising the world, and told man about it in an alphabet of the same number of letters. It is therefore no coincidence that the Revelation of St John the Divine has twenty-two chapters, and St Augustine's *City of God* was written in twenty-two books.

All of which explains why the Tarot pack has twenty-two Major Trumps known as the Greater Arcana and why occultists refer to the 'Twenty-two paths' of their travels. But it does not explain Joseph Heller's choice of the title *Catch-22* for his most famous novel, particularly in view of his decision to change it from *Catch-18* shortly before publication.

Twenty-two also has two meanings beyond the purely numerical. Since twenty-two carat used to be the quality of gold required for coinage, the expression 'twenty-two carat' or simply 'twenty-two' signified a level close to perfection. (Strictly speaking, this should be twenty-two karat, since a carat was originally a measure of weight and karat was a measure of purity, but the two are now so confused that most authorities recognise no difference between them.) Twenty-two can also mean a rifle, from the .22 calibre bullets it fired.

Twenty-two is also the number of:
- balls in snooker
- children fathered by Siamese twins Chang and Eng Bunker
- chromosomes of a hamster
- consonants in the language of the Kiowa Indians of SW America
- days in the gestation period of a rat
- different colours of undyed llama hair
- different meanings of 'fine' listed in the *OED*
- dynasties of Chinese emperors
- grand slam tennis titles won by Steffi Graf
- hours a koala sleeps each day on average
- percentage of the world's beer made in the US, cheese made in the US and wine made in France

• srutis (quarter-tones) in an octave of Indian music
• times on average an American opens the fridge each day
• yards between the wickets in cricket

Twenty-two is a rare visitor to film titles, except for these two:
• *Twenty Plus Two* (1961): detective story with David Janssen
• *Catch-22* (1970): Alan Arkin in the screen version of Joseph Heller's book

Finally, lets us not forget the findings of T. L. Kinsey, as reported in *Audio-Typing and Electric Typewriters* in 1964: '*It takes twenty-two times more mechanical energy to operate a manual typewriter than to operate an IBM electric typewriter.*'

And, as Lord Byron put it in his 'Stanza Written on the Road Between Florence and Pisa': '*The myrtle and ivy of sweet two-and-twenty Are worth all your laurels, though ever so plenty.*'

23

'She was married, charming, chaste and twenty-three.'
Lord Byron, *Don Juan*, 1819–24

Twenty-three is the smallest number of people you have to have in a room for there to be a greater than even chance that two of them have the same birthday. (Forgetting the awkward case of anyone born on 29 February, which adds a messy complication to the calculation: the second person has 364 chances out of 365 of having a different birthday from the first, the third has 363 chances out of 365 of avoiding the birthdays of the first two, the fourth has 362 out of 365 chances of differing from the other three, and if you multiply 364/365 × 363/365 × 362/365 ×

361/365, and so on, the answer drops below 0.5 when the twenty-second term is reached.) That fact, however, has nothing to do with the expression 'twenty-three skidoo', meaning scram, which first appeared in the United States in the early years of the twentieth century: nobody knows where it came from. The 'skidoo' probably owes something to the older 'skedaddle' (or simply skids), but why the 'twenty-three' became attached is a mystery. This explanation, from J. F. Kelly's *Man With Grip* (1906) seems as likely as any: *'I can see a reason for "skidoo," said one, 'and for "23" also. Skidoo from skids and "23" from 23rd Street that has ferries and depots for 80 per cent of the railroads leaving New York.'*

Twenty-three is also the number of:

• camels once offered by an Arab for the actress Diana Dors
• centimetres in a breadth – a measurement in flag-making dating from a time when flag cloth was made in 23 cm strips
• hours a week spent on household chores by the average British father
• letters in the longest word occurring in a place name in the US: Nunathloogagamintbingoi Dunes, Alaska
• meteorites known to have fallen on the UK
• minutes taken for an eighteen-month-old girl to toddle the length of Pall Mall on 11 May 1749 winning several bets by seven minutes
• people on a grand jury in the US
• percentage of the world's tangerines from Japan, watermelons from China, and TV receivers made in China
• ratio of land dug up in the US for resource exploitation to that filled with garbage
• tenses in the Santali language of India
• towns and cities in the US called Moscow
• tons of strawberries eaten at Wimbledon in 1990
• years Yorick's skull was buried before Hamlet came across it. ('This skull hath lain i' th' earth three-and-twenty years.')

In films, and other works, twenty-three is memorable for:

- *Pier 23* (1951): detective suspense
- *23 Paces to Baker Street* (1956): Van Johnson is a blind writer who foils a kidnapping
- Poem: 'On His Being Arrived at the Age of Twenty-Three' by John Milton. ('How soon hath Time, the subtle thief of youth, Stolen on his wing my three and twentieth year . . .')

24

'This is the kind of Babylonish lexicography of Johnson's Dictionary, which gives twenty-four meanings, or shadows of meaning, to the word "from".'
J. Gilchrist, 1816

Twenty-four is a number of great significance in the Jain religion and in the Isle of Man. In the latter, it is the number of members of the legislative assembly the 'House of Keys' which, from 1585 to 1734, was officially termed 'The Twenty-four Keys'. In the Manx language, it is commonly referred to as *'Yn Kiare as Feed'* (the four and twenty).

The Jain religion is said to have had 24 founding prophets, the *Jinas*, also known as *Tirthankara*, who can be distinguished from each other only by their colour, stature and longevity. Two are red, two white, two blue, two black and sixteen golden, or yellowish-brown. On the other criteria, they vary from *Rishaba*, the first *Jina*, who was 500 poles tall and lived 8,400,000 great years, to *Jahávina*, the last of *Jina*, who was the size of an ordinary man and lived on earth for only forty years.

Mathematically, a fluoroid is, according to the *OED*, a solid bounded by twenty-four triangular planes which plays an important role in crystallography. That description could fit either the tetrakis-hexahedron or the triakis-octahedron. These should not be confused with the icositetrahedron, a figure bounded by twenty-four equal and similar trapezia.

Twenty-four is also the number of:

- blackbirds baked in a pie
- carats (or, to be more precise, karats →22) in pure gold
- dollars paid to the American Indians for Manhattan Island in 1626
- draughtsmen in a set
- grains in a pennyweight
- hours in a day
- letters in the Greek alphabet
- letters in the Korean alphabet
- points on a backgammon board
- recorded accidents in the UK in 1994 involving toilet rolls
- ribs of a human (7 true pairs, 5 'false' pairs)
- scruples in a Troy ounce
- string quartets by Mozart
- syllables in a sijo – a Korean lyric poem of three lines

In film titles, the number '24' is nearly always followed by the word 'Hours', as the following list testifies:

- *24 Hours* (1931): Regis Toomey murders Miriam Hopkins, but Clive Brook is convicted (based on the novel by Louis Bromfield)
- *24 Hours in a Woman's Life* (1961): widow Ingrid Bergman falls in love with a gambler
- *24 Hours to Kill* (1966): Mickey Rooney crime caper
- *24 Hours in a Woman's Life* (1968): romance drama based on a story by Stefan Zweig
- *24 Hours to Midnight* (1992): martial arts gangster story

Literature includes:

- *The Twenty-Four Days Before Christmas* (1964) by the American children's author Madeleine L'Engle

25

'Touchstone: How old are you, friend?
William: Five and twenty, sir.
Touchstone: A ripe age.'
William Shakespeare, *As You Like It*, 1600

Twenty-five, said Beverley Nichols in his book *Twenty-Five*, is the ideal age at which to write your autobiography. It is, of course, also the square of the hypotenuse in the smallest Pythagorean triangle with sides of integer length. It is also the only square that is two less than a cube: $25 = 5^2 = 3^3 - 2$.

Dylan Thomas chose this number for his collection called *Twenty-Five Poems*, and there is a 25-yard line in hockey and formerly on a rugby pitch. In Cockney slang, twenty-five pounds is called a pony, though nobody knows why. Two ponies, of course, make a monkey (→50).

Twenty-five is also the number of:

- appearances by Richard Cohen in the British Sabre championships finals in consecutive years, the record figure for any publisher of a Book of Numbers
- beats per minute of an elephant's heart
- ducks in test matches scored by the West Indian Courtney Walsh
- gallons of soft drink drunk per capita each year in Australia
- helicopter pads in Antarctica
- hours a week the average married couple in the UK spend watching television
- minutes an average redheaded woman spends on a session of lovemaking, according to a 1999 study in Hamburg
- minutes a week the average married couple in the UK spend kissing
- minutes it takes the average Briton to get to work
- moles on an average adult body
- miles per hour of the speed limit in Britain in 1903

- percentage of the world's asses in China
- seconds you must wait on average for a new Japanese baby to be born
- years' marriage for a silver anniversary

Films with '25' in the title include:

- *25, Firemen's Street* (1973): drama of wartime upheaval in Budapest
- *The 25th Hour* (1967): World War II drama with Anthony Quinn

26

'It was not long after that that everybody was 26. During the next two or three years all the young men were 26 years old. It was the right age apparently for that time and place.'
Gertrude Stein, *The Autobiography of Alice B. Toklas*, 1933

Twenty-six is the number of letters of the alphabet and the Counties of Ireland that formed the Irish Free State in 1921: Galway, Leitrim, Mayo, Roscommon, Sligo, Carlow, Dublin, Kildare, Kilkenny, Laoighis, Longford, Louth, Meath, Offaly, Westmeath, Wexford, Wicklow, Clare, Cork, Kerry, Limerick, Tipperary, Cavan, Donegal, Monaghan.

The 26 letters of the English alphabet have by no means such a clear origin. The earliest writing in the British Isles dates back to around the third century and used a basic runic alphabet (called 'futhork' or 'futhark') of 24 letters, which was in common use in Europe at the time, together with other letters added to cope with sounds in various Anglo-Saxon dialects. Gradually this was supplanted by the Roman alphabet, enhanced with additional letters to represent sounds such as w, th (as in bath) and th (as in that). These last two letters were known as 'thorn' and 'eth', but were not used in all parts of the country. The modern 26-letter alphabet only became fully established throughout the country with the publication of Dr Johnson's dictionary in 1755.

Twenty-six is also the number of:
• black or red cards in a pack
• chromosomes of a frog
• French people expelled from Britain between 1907 and 1914 for prostitution and procuring
• languages the game of Monopoly has been translated into
• metres thickness of the walls of ancient Babylon
• New Zealand's Test Match score against England 1954–55, the lowest ever by a Test side
• percentage of the world's strawberries grown in the US
• pictures of Princess Diana submitted in the legal application to copyright the image of her face
• popes to have been assassinated
• self-governing states of India
• species of goat
• times Mount Mayon, a volcano on Luzon in the Philippines, erupted on 24 March 1993
• vertebrae in the human spine
• ways Giles Rose, chef to Charles II, could fold table napkins

Twenty-six is also where the numbers stop in the *OED*: 'One' to 'twenty-five' are all listed as headwords with meanings of their own, but twenty-six is listed only in the phrase 'Twenty-six Counties'.

Twenty-six does, however, appear in the title of one film, one book and one revolutionary movement:
• *26 Men* (1958): a made-for-TV western
• *Twenty-Six Men and a Girl*, by Maxim Gorky
• the 'Twenty-sixth of July Movement', Fidel Castro's group of Cubans that overthrew the dictator Fulgencio Batista in 1959

27

'Volumnia: He had before this last expedition twenty-five wounds upon him.
Menenius: Now it's twenty-seven; every gash was an enemy's grave.'
William Shakespeare, *Coriolanus*, 1607

Twenty-seven, apart from being the cube of three, has a number of fine mathematical properties. First, it has been proved that every integer is the sum of at most twenty-seven prime numbers. Second, if you take any three-digit multiple of twenty-seven and move the last digit to the front, or the front digit to the end, then the new number formed will also be divisible by twenty-seven. For example: $783 = 27\times29$; $837 = 27\times31$; $378 = 27\times14$.

Finally, there is a game with numbers called the 'Syracuse algorithm': start with any number; if it is even, halve it; if it is odd, multiply it by three and add one. Then repeat the process with the answer. If you proceed in this fashion it is believed (but has never been proven) that whatever number you start with you will always eventually get to the cycle, 4, 2, 1, 4, 2, 1 . . . The number 27, however, gives you one of the longest runs of all small numbers, starting 27, 82, 41, 124, 62, 31, 94, 47 . . . and peaking at 9,232 before eventually reaching 4, 2, 1 after 111 steps.

Twenty-seven is also the number of:

- amendments to the US constitution
- charges against the king in the US Declaration of Independence
- cubic feet in a cubic yard
- miles per hour in the maximum speed of a human runner
- minutes a day the average American spends reading a newspaper
- plastic bottles needed to make enough PVC yarn for a jumper
- players a side in the sixteenth-century Italian game of Calcio – an early form of football. Players had to be gentlemen, aged 18–45, beautiful and vigorous, of gallant bearing and good report

Two films have twenty-seven in the title:

- *The 27th Day* (1957): aliens give earthmen capsules that can destroy the world
- *27a* (1974): Australian drama

And one play:

- *Twenty-Seven Wagons Full of Cotton* by Tennessee Williams

28

'Mr Clarke packed all his furs on 28 horses.'
Washington Irving, *Astoria*, 1849

Twenty-eight centimetres per second is the top speed of a lone lobster. However, a line of lobsters, each clinging to the tail of the one in front, have been observed moving at speeds of up to thirty-five centimetres per second as they walk along the sea bed. The kinetics of such lobster migration queues have attracted some research, including one famous experiment in which lobster carcasses were tied together with metallic wire and dragged, by a weight and pulley apparatus, along the bottom of a tank. The experiments proved that lobsters reduce water resistance by linking in a line. Observers of migrating lobsters have also noticed that the front lobster in a queue will periodically stand aside, let the others pass, then join in again at the back, just as in a cycle pursuit team.

Lobsters apart, twenty-eight is an important number in Islam, with the link between the twenty-eight mansions of the moon in astrology and the twenty-eight letters of the Arabic alphabet seen as highly significant by some mediaeval theologians. Since the Koran also mentions exactly twenty-eight Prophets before Muhammad, it ties together beautifully. The twenty-eight Arabic letters were also used as the basis for a counting system with the first nine letters used as the numbers 1 to 9, the next nine letters representing the tens, from 10 to 90, and the next nine the hundreds from 100 to 900. The final letter represented 1,000. Much earlier, a similar counting system was used in the Egyptian *Thousand Songs of Thebes* written around 1300 BC which comprises just twenty-eight poems.

Mathematically, 28 is a perfect number, being equal to the sum of its divisors: 28 = 1+2+4+7+14.

Twenty-eight is also the number of:

- countries that sold arms to both sides in the Iran–Iraq war, 1980–88
- days in February in a non-leap year
- days it takes our outer layer of skin to be totally replaced
- different sports at the Sydney 2000 Olympics
- digits in a cubit in ancient Egypt: a digit was the width of a finger, a cubit was the length from elbow to fingertip
- hours a week the average American spends watching television
- inches height of cricket wicket
- inches length of the common or true civet (according to Hulme, 1862)
- inches length of the Congo snake (according to Tenney, 1865)
- letters in the Phoenician alphabet
- people per square kilometre in the US
- percentage of the world's tea from India
- points on the Beaume scale of saccharinity
- pounds record weight of a radish
- pounds weight of carbon in the average human body
- properties for sale on a Monopoly board
- weeks 'Don't be Cruel' was in the US charts in 1956. The B-side was 'Hound Dog'. The singer was Elvis Presley

And let us not forget the Twenty-eight Parakeet – a yellow-collared parakeet of Australia whose cry sounds like 'twenty-eight'.

The only film is:

• *The Unbeaten 28* (1981): American-made kung fu violence

29

'Nine-and-twenty-knights of fame
Hung their shields in Branksome Hall;
Nine-and-twenty squires of name
Brought them their steeds to bower from stall;
Nine-and-twenty yeomen tall
Waited duteous on them all:
They were all knights of mettle true,
Kinsmen to the bold Buccleuch.'
Walter Scott, *The Lay of the Last Minstrel*, 1805

For a prime number, 29 is unexceptional, except perhaps for the fact that $2n^2+29$ is prime for all values of n from 1 to 28. Even that, however, is not as interesting as the 'Song of Twenty-Nine' by Oliver Wendell Holmes, which contains the couplet: 'At last the day is ended, The tutor screws no more'. (Screw meaning to examine in great detail.)

Twenty-nine is also the number of:

• active volcanoes in the Kamchatka peninsula, Russia
• adults living in the islands of St Kilda – none of whom bothered to get his or her name on the electoral register for the 1997 British general election (twenty-two ran a missile-tracking station, six studied sheep, one was warden of a bird sanctuary)
• airports with paved runways in Azerbaijan
• bones in the skull (though some sources prefer the figure of 22, which excludes the ears and the hyoid bone, between jaw and larynx)
• days in February in a leap year
• degrees of angle of arc permitted for a javelin throw

- feet of intestine in a human body
- hours it took to get from New York to Chicago in 1873
- languages into which Scrabble has been translated
- letters in floccinaucinihilipilification, the action of estimating as worthless; a word used by both Walter Scott and Robert Southey
- provinces of the Byzantine empire
- racial groups of man according to Joseph Deniker (1900) and Egon von Eickstedt (1934), though their classification systems differed
- symbols in Palantyne, a mechanical shorthand invented in 1940
- slang terms for a hangover in Finnish
- the track for the Chattanooga Choo-Choo in the 1941 song by Mack Gordon (music by Harry Warren)
- waterless urinals at the Millennium Dome in Greenwich

Twenty-nine also crops up in the titles of the following films:

- *Track 29* (1988): psychodrama/satire written by Dennis Potter
- *29th Street* (1991): lottery-winner meets mobsters in comedy-drama based on true story

30

'Some thirty inches from my nose
The frontier of my Person goes.'
W. H. Auden, *Prologue: the Birth of Architecture*, 1966

Thirty is traditionally the age most women lie about. 'The best ten years of a woman's life are between twenty-nine and thirty' went the old saying. Indeed, a recent informal survey of the stated ages of successful women confirmed that considerably more of them were twenty-nine than might reasonably have been expected. Yet Honoré de Balzac knew the truth. In his novel *The Woman of Thirty*, he wrote in 1832: 'For a young man, a woman of thirty has irresistible attractions.'

Thirty is also the number of:

• bones in an arm, including the hand (though some authorities say there are thirty-two)
• cheeses that can be made from nine gallons of cow's milk, according to Aristotle
• classic races won by Lester Piggott
• days that September, April, June and November hath (from a rhyme by William Harrison, written in 1577)
• edges on a dodecahedron or icosahedron
• fatal accidents on British airlines since 1950
• goals scored for Scotland by both Kenny Dalglish and Denis Law
• magistrates – the Thirty Tyrants – imposed by Sparta over Athens at the end of the Peloponnesian War in 404 BC
• pieces of silver given to Judas
• points for a trick in spades, hearts or no-trumps (after the first) in bridge
• relations one could not marry before 1949 (→20 for the full list)
• upright stones originally in the Sarsen circle at Stonehenge (of which sixteen are still standing)
• variations in Bach's Goldberg Variations
• years of age an ancient Roman had to have attained to become a tribune
• years' marriage for a pearl anniversary

Films and other works with thirty in their titles include:

• *Thirty-Day Princess* (1934): comedy with Cary Grant
• *Thirty Seconds Over Tokyo* (1944): Van Johnson bombs the Japs
• *30* (1959): a drama set in Los Angeles
• *The 30-Foot Bride of Candy Rock* (1959): sci-fi comedy – Lou Costello's last film, and his only one without Bud Abbott
• *30 Years of Fun* (1963): silent movie comedy compilation

• *30 Winchester for El Diablo* (1965): western
• *30 is a Dangerous Age, Cynthia* (1968): Dudley Moore as a pianist seeking fame before he is thirty
• *Hurry Up or I'll be Thirty* (1973): Danny de Vito comedy
• *Thirty Tales* (1934), short stories by H. E. Bates
• *Thirty Years a Detective* (1884), an autobiographical reminiscence by Allan Pinkerton

31

'Toad that under cold stone
Days and nights hast thirty-one
Sweltered venom sleeping got,
Boil thou first i' the charmèd pot.'
William Shakespeare, *Macbeth*, 1606

Thirty-one is the number of days in most of our months, though the reasons date back two thousand years for their lurching between thirty and thirty-one, with February lagging behind, even in a leap year. The word 'month' has the same root as 'moon', and it means the time between one new moon and the next. Which, if adhered to strictly, would give an average of about twenty-nine-and-a-half days. The trouble, as usual with anything to do with the calendar, is that the times of the relative orbits of the earth round the sun (a year), the moon round the earth (a month), and the earth about its axis (a day) are not integral multiples of one another. The Romans began with ten months: Martius (of the god Mars), Aprilis (derivation obscure), Maius (goddess Maia), Junius (goddess Juno), Quinctilis (fifth), Sextilis (sixth), September (seventh), October (eighth), November (ninth) and December (tenth). Then Januarius (Janus) and Februarius (of Februa, the feast of purification) were added, and the fifth and sixth months were renamed after Julius and Augustus Caesar. March, May, July and October had thirty-one days, February had twenty-eight, and the rest had twenty-nine, which added up to 355 days, some ten days short of a full year. An extra month called Mercedonius was therefore added in alternate years, beginning after 22 February, and going on for twenty-two or twenty-three days before February was resumed. The result was a four-year cycle, of 355, 377, 355 and 378 days. Or it would have been if the priests had not occasionally neglected to put in the extra month.

In 46 BC, the calendar was ninety days out, so Julius Caesar asked the mathematician Sosigenes to take a hand. The result was the Julian calendar, which came into force in 45 BC. First, however, Julius Caesar had to resolve the problems of 46 BC, which afterwards became known as the 'year of confusion'. By the time it was over, 46 BC had had 445 days – the usual 355, plus a Mercedonius of twenty-three and an extra sixty-seven days, to give the sun a chance to catch up, inserted between November and December. For the rest of the story, →365.

Thirty-one is also the number of:

- bones in a human leg
- children of Orihah in the Book of Mormon
- counties in the US named after George Washington
- days in January, March, May, July, August, October, December
- days in the gestation period of a rabbit
- Grand Prix wins by Nigel Mansell
- letters in the Cyrillic alphabet, though the question of whether or not to include certain old letters which are obsolete in modern Russian, and whether you count accented and unaccented versions of the same letter as the same or different, makes it possible to justify almost any total from 30 to 35
- marriages in Sweden in 1990 between female doctors and male nurses (→2,396)
- mean February temperature in Denmark in degrees Fahrenheit
- miles length of the Channel tunnel
- people who died (in the short term) from radiation sickness or burns

after the accident at Chernobyl in 1986
- points the played cards may not exceed in a hand of cribbage
- states in Mexico
- syllables in a Tanka, an ancient Japanese verse form once restricted to the imperial family
- vehicles per kilometre of road in the US
- years Josiah reigned in Jerusalem (Second Book of Kings)

Thirty-one is very rare in film titles, with the exception of:
- *Adalen 31* (1969): Swedish romance directed by Bo Widerberg

'Thirty-one' is also the name of a card game in which the object is to attain a score of exactly thirty-one points – which explains the quotation that begins our next number.

32

'Well, was it fit for a servant to use his master so;
being, for aught I could see, two and thirty, a pip out?'
William Shakespeare, *The Taming of the Shrew*, 1594

Thirty-two is the fifth power of two and therefore the number of leaves of paper you end up with if you fold a single sheet in half five times, then cut the edges. This explains why the English language is blessed with a word as ugly as thirtytwomo (also written as 32mo or XXXIImo) – the size of a book made of sheets folded in such a fashion.

As the fifth power of two, thirty-two is also the number of subsets that can be formed from five objects. In other words, it is the number of different ways in which you can hold up the fingers of one hand.

Thirty-two is also the number of teeth in a full human set, which is why Horace Fletcher, in his *ABC of Nutrition* (1903), recommended you should chew each mouthful of food precisely thirty-two times. William

Gladstone followed this advice, though the *Practitioner*, in June 1907, scornfully commented: 'The Fletcherites, who, so far from not giving two bites to a cherry, insist on thirty-two to a mashed potato.'

Thirty-two is also the number of:

- clues in the world's first crossword (which appeared in *New York World* on 21 December 1913)
- days in office of William Harrison, the shortest-serving US President
- demi-semiquavers in a semibreve, which is why the Americans, who call the semibreve a 'whole note', refer to the demi-semiquaver as a 'thirty-second note'
- directions on a compass
- Fahrenheit degrees for the freezing point of water
- inches of mercury in normal barometric pressure
- pints' capacity of a pig's stomach

Thirty-two is the smallest number not to have occurred in a film title.

33

'When think you that the sword goes up again?
Never, till Caesar's three and thirty wounds
Be well avenged.'
William Shakespeare, *Julius Caesar*, 1599

'Thirty-three' was the term once used for a long-playing gramophone record, though its rate of rotation was actually thirty-three and a third revolutions per minute. First announced in 1948, and introduced in 1951, the 'long-playing' records could contain twenty minutes of music, compared with just over four minutes on the old seventy-eight rpm records. The amount of music possible on one record doubled again in 1956, when 'fine-groove' recording equipment was introduced. Then compact discs arrived, and we threw all our gramophone records, 33s, 45s and 78s, away. Anyone born in '33, it should be noted, will have been 45 in '78.

Thirty-three also plays a strong tune in Judaeo-Christian tradition, with King David reigning for thirty-three years and Jesus Christ living on earth for thirty-three years.

Thirty-three is also the number of:

- cantos favoured by Dante in the Divine Comedy
- counties in Scotland until 1975, when the counties were abolished and amalgamated into nine administrative regions, which were in turn replaced in 1996 by thirty-two unitary authorities
- days' record for going round the world by car
- Diabelli variations by Beethoven
- films made by Elvis Presley
- French expelled from Great Britain between 1907 and 1914 for brothel-keeping
- 'The Immortal Thirty-three', a group of Uruguayan patriots who, in 1825, revolted against Brazilian rule, leading eventually to the establishment of Uruguay as an independent republic in 1830
- inches height of the barriers in women's 100m hurdles
- islands in Kiribati
- kilometres per hour average speed on a London tube train, including stops
- kilos of pasta the average Italian eats each year
- life expectancy in Britain in the Middle Ages, though high infant mortality rates pulled this figure down. Most people who survived infancy lived into their forties
- percentage of world's coconuts grown in India
- pounds of rabbit meat eaten each year by the average person living in Naples
- species of parrot in the world
- species of Terrapin listed in the Catalogue of Animals at the London

Zoological Gardens in 1896
• terms of abuse for tax collectors listed in the *Onomasticon* of the Greek grammarian Julius Pollux (180–238)
• warships in Napoleon's fleet at the Battle of Trafalgar, which were destroyed by the twenty-seven ships under Nelson's command.

Thirty-three is a better number for film titles than its two predecessors, having appeared in:
• *Case 33: Antwerp* (1965): spy story
• *Naked Gun 33⅓: The Final Insult* (1994): cop spoof with Leslie Nielsen

Literature includes:
• *Thirty and Three* (1954), a collection of essays by the Canadian author Hugh MacLennan.

34

'Stella this day is thirty-four
(We shan't dispute a year or more.)'
Jonathan Swift, *Stella's Birthday Works*, 1718–19

Thirty-four is an interesting number in the study of thermometers, for it is the average Celsius temperature at the hottest place on earth, which is Dallol, Ethiopia, as measured between 1960 and 1966. It is also the average Celsius temperature of the hedgehog, which sounds potentially very uncomfortable for Ethiopian hedgehogs.

Thirty-four is also the number of:
• consonants in the Siamese language (to go with its twelve vowels), according to the traveller H. Malcom writing in 1840
• facial muscles used in vigorous kissing
• miles between West Ruislip and Epping, the longest journey you can

make on the London underground without changing
• miles that must be walked at a normal brisk pace to get rid of one pound of fat
• percentage of the world's olive oil from Italy, oranges from Brazil and gold mined in South Africa
• percentage of Anglican priests who can recite the ten Commandments correctly
• Test-match centuries of the Indian batsman Sunil Gavaskar
• years age of the oldest cat on record

35

'No person except a natural-born citizen, or a citizen of the United States, at the time of the adoption of this Constitution, shall be eligible to the office of President; neither shall any person be eligible to that office who shall not have attained to the age of thirty-five years, and been fourteen years a resident within the United States.'
US Constitution

Thirty-five is an infuriating number for makers of mathematical games and puzzles. It is the number of distinct hexominoes – shapes formed from six squares joined by their edges – and the total area of those thirty-five shapes is therefore 6×35, which is 210 square units. Yet although 210 is the area of a variety of rectangles (3×70, or 5×42, or 6×35, or 7×30, or 10×21, or 14×15) which might plausibly be filled with the hexominoes, there is, in fact, no way of putting the thirty-five shapes together to form any of them.

Thirty-five is also the number of:
• articles of clothing bought by the average American male each year
• blends of tea in the average teabag
• cubic feet of gas a cow belches each day

• days in the gestation period of a hedgehog
• days on which Easter may fall (according to a parliamentary act in the reign of George II, Easter falls on the first Sunday after the full moon which happens on, or next after, 21 March. In 1928 the House of Commons agreed to fix Easter in the week 9–15 April, subject to satisfactory consultations among the various Christian Churches. Those consultations have yet to be concluded.)
• dots on a computer screen needed for all the letters of the alphabet (→576)
• feet in length of the longest reticulated python ever measured
• fox ranches in Alaska in 1900
• gallons of water a camel can drink in 10 minutes
• metres length of a standard toilet roll
• muscles used to move the human hand
• Russians and Poles expelled from Great Britain between 1907 and 1914 for brothel-keeping
• seconds it took an enormous angry rhinoceros to eat up James Henry Trotter's parents in Roald Dahl's *James and the Giant Peach*
• tribes of the Roman people, from whom a body of judges, three from each tribe, was appointed to decide civil disputes. For the sake of linguistic simplicity, this body of 105 men was called the *centumvir* (hundred men)
• years' marriage for a coral anniversary

Books:
• *Thirty-Five Years of a Dramatic Author's Life* (1859) by E. Fitzball
• *Thirty-Five Poems* (1940) by Sir Herbert Read
• *Tall, Balding, Thirty-Five* (1966) by Anthony Firth

36

'A courtmartial sat upon him, and he was asked which he liked better, to run the gauntlet six and thirty times through the whole regiment, or to have his brains blown out with a dozen musket balls?'
Voltaire, *Candide*, 1759

The first Chinese Empire, or so the legend goes, was divided into thirty-six provinces and surrounded by thirty-six foreign peoples. As a square of the first perfect number, six, thirty-six is the sort of number to crop up in legend and mysticism. So perhaps it is no surprise that the three aspects of each of the twelve zodiac signs give a total of thirty-six.

Thirty-six is the number of Stratagems of Shanshiliu Ji, a Chinese philosopher who flourished sometime before AD 500. They are:

- STRATAGEMS WHEN WINNING

1 Crossing the sea by treachery
2 Besiege Wei to rescue Zhao
3 Murder with a borrowed knife
4 Let the enemy make the first move
5 Loot a burning house
6 Feint to the east, attack to the west.

- STRATAGEMS FOR CONFRONTING THE ENEMY

7 Make something out of nothing
8 Secretly cross at Chencang
9 Watch the fire from the opposite shore
10 Hide a dagger with a smile
11 Lead away a goat in passing
12 Sacrifice plums for peaches

- STRATAGEMS FOR ATTACK

13 Beat the grass to startle the snake

14 Reincarnation
15 Lure the tiger out of the mountains
16 Allow the enemy some latitude so you can finish him off later
17 Throw out a brick to attract Jade
18 To catch bandits capture their leader

• STRATAGEMS FOR CHAOTIC SITUATIONS
19 Pull the firewood from under the cauldron
20 Fishing in troubled waters
21 The cicada sheds its skin
22 Close the door to capture the thief
23 Make friends with distant countries and attack your neighbour
24 Borrow a road to send an expedition against Guo

• STRATAGEMS FOR GAINING GROUND
25 Replace the beams and pillars with rotten timber
26 Point at the mulberry to curse the locust tree
27 Play the fool
28 Remove the ladder after the enemy goes upstairs
29 Put fake blossoms on the tree
30 The guest plays the host

• STRATAGEMS WHEN LOSING
31 The beauty trap
32 The empty fort ploy
33 Counter-espionage
34 The self-injury ploy
35 Interlocking stratagems
36 Sometimes retreat is the best option

Thirty-six is also the number of:

- feet width of a tennis court (for a doubles match)
- grams of buns, scones and teacakes eaten by the average Briton each week
- grams of butter eaten by the average Briton each week

• inches height of hurdles in the men's 400m event
• inches in a yard
• miles travelled each year by taxi by the average Briton
• percentage of Britons who made New Year's resolutions in 1999
• years' life expectancy at birth in Malawi, the world's lowest

In the cinema, thirty-six is celebrated by the following films:
• *36 Hours* (1965): World War II suspense
• *36 Hours to Hell* (1977): World War II epic
• *36 Fillette* (1988): sexual awakening of 14-year-old French girl – the title is a well-known dress size

37

"Women should marry when they are about eighteen years of age, and men at seven and thirty.'
Aristotle, *Politics*, *c.* 330 BC

Thirty-seven is a significant number in the history of flying. It was the length in metres of the first flight made by Orville Wright on 17 December 1903, and was also the number of minutes it took Louis Blériot, on 25 July 1909, to become the first person to fly across the English Channel.

Thirty-seven is also the number of:
• comets discovered by Jean-Louis Pons (1761–1831)
• days it took Elton John's 'Candle in the Wind' (Diana Memorial version) to overtake Bing Crosby's 'White Christmas' as the best-selling single of all time
• degrees Celsius in normal human body temperature
• executions in Texas in 1997
• grams of protein needed each day by a man of average weight

• metres per minute at which escalators in underground stations and airports travel. Those in shops tend to go at the slower speed of 27 metres per minute.
• plays of Shakespeare
• taxi drivers in New York city named Amarjit Singh
• US states with the death penalty
• universities in Australia
• volumes in the encyclopaedia of natural history compiled by Pliny the Elder

38

'Give me a thirty-eight every time. Just flick back the hammer and let her go. I'll drop anyone at five hundred feet.'
William Burroughs, *Junky*, 1953

Thirty-eight is a number associated with danger. Quite apart from being the calibre of a .38 automatic, it is the number of people in the United Kingdom who died of Creutzfeldt-Jacob Disease in 1995 and the number of recorded accidents in homes in the UK in 1994 involving Christmas trees.

Thirty-eight is also the number of:
• aeroplanes on show at the world's first great aviation meeting near Rheims, France, in 1909
• bathrooms in the house that Mike Tyson put up for sale in 1997
• chromosomes of a cat
• inches in the maximum length of a cricket bat
• miles a year cycled by the average Briton
• minutes' length of the war in 1896 between Britain and Zanzibar
• percentage of Canadian women who prefer chocolate to sex, according to a survey in 1996

• professional teams in the Scottish Football League
• streets named 'Vostochnaya' (eastern) in the Ukrainian city of Kherson in 1995, when a special renaming committee was established to ease the confusion
• times the England football manager Graham Taylor said 'fuck' in the TV documentary 'An Impossible Job' in 1995

Film:
• *38: Vienna Before the Fall* (1988): cross-cultural romance under the Nazis

39

'Then I gave not over opening place after place until nine and thirty days were passed, and in that time I had entered every chamber except that one whose door the Princesses had charged me not to open.'
Sir Richard Burton, *Tales from the Arabian Nights*, 1888

The main significance of thirty-nine is that it is one less than forty, but forty, as we shall see, is a number of such great importance that being one less than it is quite an honour. If forty is the upper limit ascribed to something, then thirty-nine is as far as you are allowed to go. So when forty lashes was considered the greatest permissible punishment, the Law of Moses prescribed a maximum of thirty-nine. St Paul, in his letter to the Corinthians, refers to himself receiving 'forty blows less one'. At about the same time there was drawn up a list of the thirty-nine principal types of task that were forbidden on the Jewish sabbath. Later another list was added of thirty-nine further minor things to be avoided. In both cases, the lists gave the impression of leaving one place free, to suggest that the list was not quite all-embracing.

Thirty-nine is also the number of:

• Articles of the Church of England, set out in 1563 and approved by Parliament in 1571 to settle religious disputes after the Reformation
• grams of tea consumed by the average Briton each week
• handkerchiefs used to mop the brow of George IV during his coronation in 1821
• hockey riots in the US, between 1960 and 1972
• inches height of the hurdles in the men's 110 metres event
• feet in the length of a badminton singles court
• people killed by stray bullets in NY city in 1989
• people shot dead in gun-related crimes in Japan in 1994
• percentage of world's raspberries in 1991 that came from the USSR
• Russians and Poles expelled from Great Britain between 1907 and 1914 for soliciting and importuning
• years' reign of Henry VI

But for John Buchan, we might never have seen '39' in a film title:

• *The 39 Steps* (1935): Robert Donat as Richard Hannay in Hitchcock's classic version of the John Buchan novel. Despite the general brilliance of the film, it almost omitted to mention the thirty-nine steps of the title, an explanation only being included close to the end
• *The 39 Steps* (1960): Kenneth More plays Hannay in the least of the screen versions of the book
• *The 39 Steps* (1979): with Robert Powell and Sir John Mills

40

'Nobody loves a fairy when she's forty.'
Arthur W. D. Henley, song title, 1934

Forty, apart from being the only number-word in the English language that has its letters in alphabetical order, is a number with deep Biblical significance. Particularly in the Old Testament, whenever a number is needed that is full of weight and portent, forty seems to be the natural choice. There were the forty days and forty nights during which rain fell during the Flood, then another forty days passed before Noah opened the window of the ark; Moses spent forty days on the mount; Elijah was fed for forty days in the wilderness by ravens; Nineveh was given forty days by Jonah to repent. In the New Testament, Jesus fasted for forty days, and he was seen forty days after his resurrection. Following this tradition, forty days became enshrined in law whenever a suitable period of waiting was required; hence the forty days of quarantine (from the French for 'forty') and the forty days, after the prorogation of the House of Commons, that an MP used to be free from the possibility of being arrested.

Forty is also the number of:
• atoms in a molecule of penicillin
• chromosomes of a mouse
• days a widow was allowed, under old English law, to remain in her husband's house after his death
• days allowed in old English law to pay a fine for manslaughter
• days in the gestation period of a polecat
• days of rain that will follow, proverbially, if it rains on St Swithin's Day; weather statistics, however, do not bear this out
• the 'Forty Immortals', the members of the French Academy who, once elected, are members for life
• murders in Belgrade in the first four months of 1997
• points for the first no-trump trick in a bridge contract
• qualification age in order to be fortified by Phyllosan
• thieves encountered by Ali Baba
• winks for a short nap
• years' marriage for a ruby anniversary
• years of Israel in the wilderness
• Years On, for the Harrow school song

Films with '40' include:

• *40 Pounds of Trouble* (1963): Stubby Kaye, Tony Curtis, Phil Silvers and a mischievous child
• *40 Guns to Apache Pass* (1967): western with Audie Murphy
• *40 Graves for 40 Guns* (1971): American outlaws steal Mexican relic

41

'Lizzie Borden took an axe
And gave her mother forty whacks;
When she saw what she had done
She gave her father forty-one.'
Anonymous, after the trial of Lizzie Borden in America in 1893

Never trust a homicidal daughter with an axe, or an encyclopaedia that tells you that Mozart wrote forty-one symphonies. Everyone knows, of course, that Mozart's last symphony, number 41 in C major, is the Jupiter, but some of those numbered 1 to 40 were later discovered probably not to have been written by Mozart at all, others are not symphonies but concert overtures, while several of his other compositions deserve the designation of 'symphony' although they are not known as such.

The last Symphony, incidentally, was not known as the Jupiter until some sixty years after Mozart's death. The name was bestowed upon it by Johann Baptist Cramer (1771–1858), a composer and music publisher, who thought it deserved a name that reflected its grandeur. In Germany, they call it the *Synphonie mit der Schlussfuge* (symphony with the final fugue), which is also incorrect, since the final movement, while containing some intricate counterpoint, is not, strictly speaking, a fugue.

Lizzie Borden, by the way, was acquitted of the charge of murdering her parents. All of which adds up to forty-one being one of the most consistently wrong numbers in history.

Forty-one is also the number of:

• days Phil Stubbs and Rob Hamill took to row across the Atlantic in October and November 1997
• grams of margarine eaten by the average Briton each week
• members on the Venetian council, known as the forty-one, by whom the Doge was elected
• pounds in royalties earned by Freud for his book *The Interpretation of Dreams*
• riding clubs in Iceland in 1978
• years in the longest recorded lifespan of a goldfish

42

'And in this borrowed likeness of shrunk death
Thou shalt continue two-and-forty hours,
And then awake as from a pleasant sleep.'
William Shakespeare, *Romeo and Juliet*, 1595

Forty-two is, as all fans of Douglas Adams's *Hitch-hiker's Guide to the Galaxy* know, the answer to Life, the Universe and Everything. The curious thing, however, is that while Douglas Adams claims to have chosen the number for no particular reason, forty-two also crops up in several ancient religions as a number of great significance. In ancient Egypt, the fate of the dead was supposed to be decided by forty-two demons, each representing one diocese of the country and ready to seize the soul of any individual who had committed a sin within its area of responsibility.

'I know thee, I know thy name, I know the names of the Forty-two Gods who live with thee in this Hall of Maati, who live by keeping ward over sinners, and who feed upon their blood on the day when the consciences of men are reckoned up in the presence of the god Un-Nefer.'

Egyptian *Book of the Dead*

There was a forty-two-armed Hindu god and forty-two was a sacred number in Tibet. In Judaeo-Christian tradition also, the number forty-two crops up more often than it ought. There were forty-two generations from Abraham to Jesus Christ, forty-two Levitical cities, forty-two boys torn to pieces by bears because they had ridiculed the prophet Elisha (2nd Book of Kings), forty-two sacrifices of Balach in the Book of Numbers and 'forty and two months' which the Gentiles would tread the Holy City, as predicted in the Book of Revelation. (Incidentally, forty-two months is three-and-a-half years, and if you multiply 3½ by 42, you get 147, which is the mystical number of the snooker table.)

Another writer who took a fancy to the number forty-two was Lewis Carroll, but in his case the choice was probably more deliberate. Carroll, in real life the Reverend Charles Lutwidge Dodgson, took a keen professional interest in comparative religion. In *Alice in Wonderland*, we have 'Rule 42: All persons more than a mile high to leave the Court.' Another Rule 42 crops up in the preface to *The Hunting of the Snark*, and in the same work we may read, of The Baker:

'He had forty-two boxes, all carefully packed,
With his name clearly painted on each;
But, since he omitted to mention the fact,
They were all left behind on the beach.'

Furthermore, the story-teller in *Phantasmagoria* gives his age as forty-two, although Carroll was only in his thirties when he wrote it. Finally, the price on the Hatter's hat in Tenniel's illustration of the mad tea-party is ten shillings and sixpence, which is 126 pence, and 126 is three times forty-two. There are, incidentally, forty-two illustrations by Tenniel altogether in Carroll's works.

Forty-two is also the number of:

- Articles of Religion of the Church of England drawn up in 1553, which later formed the basis for the Thirty-Nine Articles
- aspects of personality revealed by head bumps according to the pseudo-science of phrenology

• books on stamps or stamp-collecting published in the UK in 1996
• degrees of angle between the elevation of a rainbow and the observer's shadow
• different impressions left by tyres with which Sherlock Holmes claimed to be familiar in *Priory School* by Arthur Conan Doyle
• eyes on the court cards in a pack of cards
• gestation period in days of a ferret
• governments that established the League of Nations in 1918
• inches in the maximum permitted length of a baseball bat
• kilometres from Dover to Calais
• kilometres length of the Berlin Wall
• known species of the Bird of Paradise (of which you may find thirty-three in Papua New Guinea)
• lines in a 'length' – theatrical slang for a portion of an actor's part equal to forty-two lines
• lines of type in each column of most pages of the Gutenberg Bible, which is therefore sometimes called the 'forty-two-line Bible'
• metres height of Sweden's tallest Christmas tree
• miles per hour for the maximum speed of a grey fox
• minutes an object would take to fall when dropped into a straight frictionless tunnel bored through the earth. Curiously, it does not matter if the hole goes right through the centre of the earth and out the other side, or is bored at an incline to emerge wherever you choose, the effect of gravity guarantees that it will always take the same forty-two minutes to come out the other end
• percentage of the London underground network that is underground
• provinces of Ancient Egypt, called 'nomes'
• staff manning the year-round stations in Antarctica
• stations that the children of Israel had to pass through between Egypt and Sinai
• teeth of a dog or wolf
• teeth of a whale, according to Herman Melville in *Moby Dick*
• wells that can be drilled from a single platform of an oil rig

• years the Holy Grail is said to have fed Joseph of Arimathea when he was imprisoned by the Romans

Forty-two is also celebrated in film:

• *42nd Street* (1933); Busby Berkeley classic musical

• *The 42nd Street Cavalry* (1974): Dennis Weaver in a mystery romance

And book:

• *Forty-Two Years Amongst Indians and Eskimo* by J. Hordern (1893)

43

'She may very well pass for forty-three
In the dusk with a light behind her.'
W. S. Gilbert, *Trial By Jury*, 1875

Life may begin at forty, but forty-three seems to be the age at which it starts to deteriorate. Apart from Gilbert's snide line, we also have Robert Browning's *'I am forty-three years old; in prime of life, perfection of estate'* (*Red Cotton Nightcap Country*, 1873), and C. M. Yonge's *'She looked her full forty-three years'* (*Cameos*, 1979).

Forty-three is also the number of:

• beans alleged to be in each cup of Nescafé

• city trades depicted on the windows of Chartres Cathedral

• pairs of nerves joining the central nervous system with rest of body

• years the Empire State Building was the tallest in the world (1931–74, at the end of which it was overtaken by Sears Tower, Chicago)

44

'Again I was lucky with the Psalms; the Sunday before there had been forty-four verses; this Sunday there were forty-three, seven below the danger line.'
L. P. Hartley, *The Go-Between*, 1953

The *Oxford English Dictionary* defines a 'forty-four' as either: a) a forty-four-gun ship; or b) a bicycle with a wheel forty-four inches in diameter. This seems a peculiarly unfortunate linguistic coincidence, full of opportunities for confusion and misunderstanding.

Forty-four is also the number of:
• feet length of a badminton doubles court
• feet length of the oars in galleys at the end of the eighteenth century
• languages into which Bram Stoker's *Dracula* has been translated
• times Britain was judged to have violated the Human Rights Convention in the period 1960–96
• years to which the oldest Tower of London raven, Jim Crow, lived

45

'Mrs Deborah no sooner observed this than she fell to squeezing and kissing, with as great raptures as sometimes inspire the sage dame of forty and five towards a youthful and vigorous bridegroom.'
Henry Fielding, *Tom Jones*, 1749

On the other hand, we have:
'A maiden of forty-five, exceedingly starched, vain and ridiculous.'
Tobias Smollett, *Humphrey Clinker*, 1771

A forty-five may be an extended-play record or the most popular hand-gun in the Wild West. The Colt .45 was not made until more than thirty years after the death of Samuel Colt, inventor, in 1835, of the revolving pistol. He became the most successful gunsmith in America when, in 1847, he received an order to supply 1,000 guns to the American army. Cartridge-loading revolvers first appeared in 1857, five years before Samuel Colt died. The first self-loading semi-automatic pistol arrived in 1895, and two years later, John Browning patented the automatic pistol which became the basis for the Colt .45.

Forty-five is also the number of:

• bells in the carillon in the city of Mechelen, Belgium
• centimetres width of sleeping space allowed to convicts on board ships taking them from England to Australia after 1802
• centimetres length of an aardvark's tongue
• degrees in the smaller angles of a right-angled isosceles triangle
• dinosaurs named by Friedrich von Huene
• grams of honey made from the nectar collected by one bee in its lifetime
• inches length of the nose of great bronze Buddha at Kamakura, Japan
• letters in the longest word to be found in any English dictionary in 1966: pneumonoultramicroscopicsilicovolcanoconiosis (small particles of ash and dust – included in the Merriam-Webster's great Unabridged)
• maximum permitted run-up for pole vault in metres
• miles per hour maximum speed of an elk
• minutes a day the average American spends listening to recorded music
• species of Coccinellid beetles found in Britain
• years' marriage for a sapphire anniversary

And look out for:

• Film: *45 Fathers* (1937): comedy
• Novel: *The Forty-Five* (1848) by Alexandre Dumas
• Musical: *Forty-Five Minutes from Broadway* (1906) by George M. Cohan

46

'She's six-and-forty and I wish nothing worse to happen to any woman.'
Sir Arthur Wing Pinero, *The Second Mrs Tanqueray*, 1893

There is a beautiful little word-game to be played with the number forty-six and the King James Bible. If you count forty-six words from the beginning of Psalm 46, you will find the word 'shake' and forty-six words from the end of the same Psalm is the word 'spear'. That translation of the Bible was published in 1610, in which year Shakespeare celebrated his forty-sixth birthday; whether Shakespeare himself was involved in the translation, or whether a fellow wordsmith had hidden the birthday greeting within the Psalm on his behalf, is a matter for speculation. Now try counting up the words in the sentence you have just read and you will understand why there is a semi-colon rather than a full stop in the middle of it.

Forty-six is also the number of:

• chromosomes of a human being
• countries belonging to the International Bureau of Weights and Measures
• days before anyone noticed Matisse's 'Le Bateau' was upside down at the Museum of Modern Art in New York in 1961 (an art student spotted it on day forty-seven)
• grams maximum weight of a golf ball
• inches height of Michael Dunn (1935–73), co-star of 1965 film *Ship of Fools*
• legs of *Scolopendra gigantea*, the world's biggest centipede. Although our calling it 'centipede' (a hundred legs) is therefore a gross overestimate, we do not do nearly as badly on this matter as the Germans, who call a centipede a *Tausendfüssler* (thousand feet). Other centipedes, however, may have far more legs (→175) than *Scopolendra gigantea*

47

'And now let us go to the tomb of the Forty-Seven Ronins.'
Rudyard Kipling, *From Sea to Sea*, 1899

The Ronin were masterless Samurai of Naganeri Asano, who dramatically avenged their master's death on 14 December 1702. Their exploits are celebrated in several films (see below).

The AK 47 automatic rifle is perhaps the world's most famous submachine gun. The letters stand for *Avtomat Kalashnikov* – the Kalashnikov automatic, after the name of its inventor. In 1997, however, Mr Kalashnikov, now well into his eighties, said that he had always intended the gun to be used for the preservation of peace. He is not related to the *Song of the Merchant Kalashnikov* which was a mature work of Mikhail Lermontov, the Russian novelist who was shot and killed (though not by an AK 47) in a duel at the age of twenty-six in 1841.

Quite apart from the forty-seven Samurai and the AK forty-seven, there is reason to be in awe of the number forty-seven, particularly if you were ever a student at Pomona College in California or a fan of *Star Trek*. The cult of forty-seven began at Pomona in the early 1960s when a group of maths students began to investigate certain numbers that seemed to be turning up more frequently than they ought. Their researches suggested that of all numbers forty-seven was most likely to crop up when you didn't expect it. There were, for example, forty-seven pipes in the top row of the college organ; there were forty-seven letters in an inscription of the dedication plaque of one hall of residence which, incidentally, was completed in 1947; there were forty-seven students in Pomona's first graduation class; and to get to Pomona from the San Bernadino Freeway, drivers are advised to take exit 47.

When a Pomona graduate became a script-writer on *Star Trek*, it was quite natural, therefore, that he chose forty-seven whenever a random number was required. The series is littered with planets where all the inhabitants have been destroyed except for forty-seven survivors; or members of the crew who have been unconscious for forty-seven seconds; or rooms numbered 47.

Forty-seven is also the number of:

• bullets that killed the Mexican revolutionary Pancho Villa
• deaths from new variant CJD in the UK (up to November 1999)
• degrees of Latitude between the tropics of Cancer and Capricorn
• members of parliament arrested on 6 December 1648 for opposing the trial of Charles I
• men hanged for rape in England and Wales between 1805 and 1818
• metres height of the pedestal of the Statue of Liberty
• miracles with which Jesus is credited in the New Testament
• Mormon temples built or under construction in the world
• opus number of Beethoven's Kreutzer Sonata
• paragraphs on kiddushin (betrothal) in the Mishnah, the ancient Jewish law
• passengers killed in crashes on British airlines between 1980 and 1989
• piglets used in filming *Babe*
• sentences in the US Declaration of Independence
• people per square kilometre in Mexico
• proposition number of Pythagoras's Theorem in Euclid's Elements.
• seats in the Senate in Kazakhstan
• sounds in the Japanese language
• square kilometres of the Pitcairn Islands
• strings on a harp
• times its own weight in excess acid the indigestion cure Rolaids claims to absorb
• years a brown bear has lived in captivity
• years life expectancy in Kenya

Films:

• *47 Ronin* (1994): revenge in early eighteenth-century Japan (*Shiijushichin no Shikaku* in the original)

- *The 47 Ronin, Part 1* (1942): earlier version of the above
- *The 47 Ronin, Part 2* (1942): more of the same

Essay:

- *On Completing Forty-Seven* by Thomas Hood. Sadly Hood never did: he died in 1845 at the age of forty-six

48

'Mr Woodhouse considered eight persons at dinner together as the utmost that his nerves could bear – and here would be a ninth – and Emma apprehended that it would be a ninth very much out of humour, at not being able to come even to Hartfield for forty-eight hours, without falling in with a dinner-party.'
Jane Austen, *Emma*, 1816. (This quotation, incidentally, is the first recorded use of the compound noun 'dinner-party'.)

The 'Forty-Eight', Johann Sebastian Bach's 48 Preludes and Fugues, properly known as *Das wohltemperierte Klavier*, is one of the most influential compositions in the history of music. They were written to demonstrate the effectiveness of 'temperament' – a method of tuning the piano keyboard to enable music to be played in any key. Essentially the system involves a compromise on many of the notes, rendering them slightly out of tune so that they will not be too far out whatever key one is playing in. (It's all to do with B flat not really being the same note as A sharp.) It is this out-of-tuneness that makes music in one key sound different from another when played on a keyboard. Bach's forty-eight are, in fact, two sets of twenty-four preludes and fugues, one in each of the twelve keys, major and minor.

The versatile divisibility of forty-eight, which is equal to 2×24, or 3×16, or 4×12, or 6×8, must be held responsible for the linguistic overkill exhibited by the words hexakisoctahedron, octakishexahedron, tetrakisdodecahedron and tetrakonta-octahedron, all of which mean the same solid body bounded by forty-eight triangular planes, as is found in the crystalline formation of diamonds. Fortunately the problem may be avoided by calling it an adamantoid.

Forty-eight is also the number of:

• recorded accidents in UK homes in 1994 involving beanbags
• cards needed to play pinochle
• chromosomes of great apes
• constellations listed by Ptolemy
• days in space for South American guppies on board Salyut 5 in 1976. These were the first fish in space
• deaths from Creutzfeldt-Jacob Disease in the UK in 1996
• letters in Bernard Shaw's logical spelling alphabet
• pairs of socks received as presents by George Bush when he was vice-president
• political divisions, known as naucraries, of the ancient Athenians
• pounds weight of an elephant's heart

Every film with 'forty-eight' in the title has that number followed by the word 'hours':

• *The Forty-Eight Hour Mile* (1970): Carrie Snodgrass in a love-triangle mystery
• *48 Hours* (1944): World War II drama from a story by Graham Greene
• *48 Hours* (1982): Nick Nolte and Eddie Murphy in comedy suspense
• *48 Hours to Acapulco* (1968): gangsters and espionage
• *48 Hours to Live* (1960): nuclear scientists are kidnapped

Book:

• *Forty-Eight Days Adrift* by J. Barbour (1932)

49

'In a cavern, in a canyon
Excavating for a mine,
Dwelt a miner, forty-niner
And his daughter, Clementine.'
Percy Montrose, *Clementine*, 1884

The Forty-Niners were speculators who rushed to California after gold was discovered there in 1848. By the end of 1849, California's population had increased from about 15,000 to over 100,000. San Francisco grew from a small town into a city of 25,000 people, and the busiest port on the Pacific coast, as it became the main place of arrival for hopeful prospectors from all over the world.

Forty-nine is also the number of:

- balls in the British lottery
- instant lottery tickets sold every second in Britain
- phonetic symbols of the zhuyin zimu pronunciation alphabet, officially promulgated in 1918 by the Chinese government
- professional fights fought and won by Rocky Marciano
- years of age at which your mind is in its prime: 'The body is in its prime from thirty to five-and-thirty; the mind about forty-nine.' Aristotle, *Rhetoric*, Fourth century BC

Films include:

- *The 49th Parallel* (1941): Nazis flee through Canada trying to escape to the USA; with Raymond Massey and Michael Redgrave
- *The 49th Man* (1953): atom bombs and espionage
- *Forty-Nine Days* (1964): US Navy saves shipwrecked Soviet sailors

Books:

- *Diary of a Forty-Niner* by A. T. Jackson (1906)

- *Experiences of a Forty-Niner* by C. D. Ferguson (1888)
- *Log of a Forty-Niner* by R. L. Hale (1900)

50

'A man shouldn't fool with booze until he's fifty;
then he's a damn fool if he doesn't.'
William Faulkner, quoted in James M. Webb and
A. Wigfall Green: *William Faulkner of Oxford*, 1965

L is the Roman number for fifty and, as with all the other Roman numerals, the letter does not stand for any particular word. Even in the case of C (which could so easily stand for *centum*, a hundred), the number-letter was originally only a symbol, which eventually became identified with the letter it resembled.

Fifty is the smallest number that can be expressed in two different ways as the sum of two squares: $50 = 1^2+7^2 = 5^2+5^2$.

Fifty is also the number of:

- days in the gestation period of a mink
- days the 'khamsin' wind is expected to blow through Egypt around the month of April. A hot windstorm, its name comes from the Arabic for fifty
- floors in Canary Wharf, Europe's tallest occupied building
- height in inches of Queen Matilda, wife of William I. She is believed to have been Britain's shortest monarch
- inches in the maximum permitted height for a member of 'The Little People of America' society
- millimetres width of the average European condom when flat
- oysters Casanova recommended eating for breakfast
- pounds sterling in a 'monkey'
- United States of America, since Hawaii joined the other 49 in 1959.

Hawaii was formerly known as the Sandwich Islands, thus named by James Cook in honour of the fourth Earl of Sandwich, First Lord of the Admiralty and the first person to enjoy food between two slices of bread
• Ways to Leave Your Lover, according to the hit song by Paul Simon, of which perhaps the most memorable is 'get on the bus, Gus'
• years' marriage for a golden anniversary

'Fifty-fifty' indicates a perfect balance between alternatives. The fifties in the expression clearly relate to percentages, and since percentages are ratios rather than pure numbers we have limited their appearances in this book. For anyone in search of a percentage, however, here is a complete table, from 0 to 99. The figures quoted are in each case based on recent surveys. The sample may be assumed to be from the UK unless otherwise stated.

0% of Americans keep gloves in their cars' glove compartments
1% of people got engaged on a foreign holiday
2% of children eat sweetcorn or broccoli every day
3% of the population will only clean the kitchen as a last resort
4% of women believe instant attraction could be sparked by the wealth of a potential partner
5% of Britons are vegetarian
6% of accountants say they'd give up work if they won the lottery
7% of Britons are afraid of spiders
8% of people say their favourite paint colour is yellow
9% of Britons claim to have made love on a foreign beach
10% of people think pet rocks are idiotic
11% of Britons have more than 500 books at home
12% of married men aged 28–33 can't remember why they got married in the first place
13% of people aged 18–24 say they would sleep with their boss to achieve promotion

14% of men have taken their football team's shirt away with them on holiday
15% of American women don't like the shape of their bottoms
16% of smokers have not told their doctors they smoke
17% of women say romantic novels are their favourite reading
18% of adults believe that world peace will be achieved within the next 100 years
19% of Britons living in the US get Cadbury's chocolate sent to them
20% of people have resorted to alcohol to relieve stress
21% of working women say their bosses are their biggest cause of stress
22% of people say they would sacrifice their lives to save a partner
23% of Britons living in Spain have National Lottery tickets sent to them
24% of television viewers enjoy the adverts
25% of British women abroad regularly get packages of underwear from M&S
26% of men aged 16–34 are worried about stress
27% of people wish their first kiss had been with someone else
28% of men daydream while driving
29% of adults believe the computer was the best invention of the twentieth century
30% of people believe their health is suffering because of their work
31% of people think the police are dishonest
32% of house-buyers say a nearby post office is very important to them
33% of the world's hops grow in Germany
34% of working women earn more than their partners
35% of Britons say the clock is what they'd most like to see on waking
36% of adults feel a television is essential to their lives
37% of British births are outside marriage
38% of the population say their lives are so stressful that they have thought of work while having sex
39% of Britons think we will find intelligent life on another planet by 2010

40% of women aged 19–34 would like to smother George Clooney in yoghurt
41% of mothers worry about their children being involved in a road accident
42% of people living in England have credit cards
43% of Britons think space-travel for all will become a reality within the next 10 years
44% of women say that sexy underwear is their favourite way of pleasing a man
45% of men dislike the house they live in
46% of dog-owners would sometimes rather spend time with their dog than with other people
47% of teenage girls would get engaged just to get a big ring
48% of working women say they would choose to be a woman of leisure if they had enough money
49% of people think pollsters tell the truth
50% of American women nearly always sit with their legs crossed
51% of people will drive when they are feeling unwell
52% of teenagers are worried about spots or skin problems
53% of people think the police are good at catching criminals
54% of men find ear-kissing really erotic
55% of working women say the stress of work causes them to shout at their children
56% of Americans send Christmas cards to their cat or dog
57% of Russians think Lenin should be removed from Red Square and buried
58% of smokers are worried about the cost of cigarettes
59% of schoolchildren accuse their parents of constant nagging
60% of women would love to wear a crinoline dress
61% of adults believe in love at first sight
62% of Hell's Angels in the US are moved by poetry
63% of single males say they like their house to look nice
64% of all retail prices end in the number 9

65% of women have used food during sex
66% of newborn babies are breast-fed
67% of people have been victims of burglary, car theft or mugging
68% of German television viewers change channels when the adverts come on
69% of people think their lives are becoming more difficult
70% of people find the presence of policemen comforting
71% of parents say exams are now harder than in their day
72% of people believe politicians cannot be trusted to tell the truth
73% of Londoners say that they are experts on the underground system
74% of British women recycle stockings, as art, pipe-lagging or onion and fruit storage
75% of motorists would not use public transport to travel to work, even if public transport costs were halved
76% of homes have window locks
77% of women shy away if someone they don't know well tries to kiss them
78% of people listen to tapes, records or CDs
79% of people trust their dentists
80% of women don't feel safe going out alone at night
81% of teenage girls say they would be happy to grow old while staying single
82% of teenagers are worried about exams
83% of people know it is a criminal offence to drop litter
84% of people would like to go on a moon shuttle
85% of Americans say they would risk their lives to save their pet
86% of 7–11-year-olds refuse to eat spinach
87% of parents have at some time smacked their children for misbehaviour
88% of male smokers are unaware that smoking can lead to impotence
89% of men and women blame work for them not having enough time to meet members of the opposite sex

90% of people are happy in their current job
91% of people believe what their doctors tell them
92% of adults say they would be willing to consider having sex in bed at home
93% of 14–16-year-olds say they think about their future from time to time
94% of people feel better after they have done the washing up
95% of British people think reading is an important and positive part of their lives
96% of Americans have been to McDonald's in the past year
97% of American girls love having their necks kissed
98% of British homes have fitted carpets in the sitting room
99% of accountants say that they have suffered from Inland Revenue errors

51

'To denote as Proof Spirit that which, at the Temperature of Fifty-one Degrees by Fahrenheit's Thermometer, weighs exactly Twelve Thirteenth Parts of an equal Measure of Distilled Water.'
Act of George III, 1818

The fifty-first Psalm, known as the 'neck Psalm', which begins, in the Latin, 'Miserere mei Deus' (Have mercy upon me, O God), was the one traditionally set for an accused criminal to read if he was to save his neck by claiming 'benefit of clergy', an exemption from civil law granted to officials of the church.

Fifty-one is also the number of:
- days in the gestation period of a fox
- founder states of the United Nations in 1945
- Grand Prix wins of Alain Prost

• nursery rhymes in *Mother Goose's Melody* (1781), including 'Ding Dong, Bell' and 'Little Tommy Tucker'
• razor sets sold by Gillette in their first year of business
• years age of the oldest known orang-utan, Julia, who died in November 1992 in Rotterdam Blijdorp zoo

52

'She has as many diseases as two and fifty horses.'
William Shakespeare, *Taming of the Shrew*, 1623
'Quoth she "Here's but two and fifty hairs on your chin, and one of them is white".'
William Shakespeare, *Troilus and Cressida*, 1602

Fifty-two, apart from being a favourite number of Shakespeare's, was also important in the calendar of the Mayan civilisation. They had a curious system of counting the passage of time, with one cycle of thirteen days operating concurrently with another cycle of twenty days. So if both cycles start together, it takes 260 days to get back to the beginning. And since the first multiple of 260 that is also a multiple of 365 is 18,980 (= 73×260 = 52×365), it takes fifty-two 365-day years for the Mayan calendar to get back to where it started in normal earth-years. The Mayans knew this, of course, and celebrated the beginning of a new life-cycle every fifty-two years.

Fifty-two is also the number of:
• British lords reputedly kept happy by Lindi St Clair (also known as 'Miss Whiplash') in twenty years of business
• cards in a pack
• Eskimo dogs that pulled Roald Amundsen's supplies on his trek to the South Pole in 1911. Only eleven of the dogs returned. The explorers had shot and eaten the weakest ones when they were no longer needed

• 'fundamental playing errors' Charles Darrow was told Monopoly had when he submitted it to Parker Brothers in 1934
• inches above the floor of the top rope around a boxing ring
• murders in the film *Natural Born Killers*
• times the word 'fruit' occurs in the New Testament
• varieties of condom available in Norway in 1991
• weeks in a year

Films include:
• *52nd Street* (1937): musical, not nearly as good or successful as the one ten blocks down (→42)
• *52 Pick-Up* (1986): Ann-Margret in a tale of violence based on a book by Elmore Leonard

53

'And the sun went down, and the stars came out far over the summer sea,
But never a moment ceased the fight of the one and the fifty-three.'
Alfred Tennyson, *The Revenge*, 1880

The poem is a romantic account of the final battle of Sir Richard Grenville (1541?–91) whose ship the *Revenge* was isolated off Flores and engaged in a fifteen-hour battle with a large number of Spanish vessels before being finally overcome. The number 'fifty-three' is an honest enough count of the Spanish force, though Tennyson does not mention that all but twenty of them were supply vessels, and only fifteen of the warships joined in against the *Revenge*. Nor does he mention that the whole battle only came about because of Grenville's stubbornness and bad temper in insisting on not letting the Spanish get in his way, rather than steering away from them.

In the 1890s the game of poker was popularly played with fifty-three cards (they called it Fifty-three Deck Poker), the fifty-third card being the joker or a blank card usually included as a spare by the makers of packs in America. It has been suggested that the original of the game lay in the meanness of people who did not want to see the blank card wasted. The name given to the blank card was 'mistigris', from an old French card-playing term which usually signified the jack of spades.

Fifty-three is also the number of:

- countries in the Commonwealth
- countries in which the Spice Girls had a number one hit within six months of their debut
- Dickin medals awarded to animals in World War II
- honorary degrees awarded to Bob Hope
- metres height of the tallest Christmas tree ever recorded
- the most miles ridden backwards on a unicycle
- the number painted on Herbie the VW in the film *The Love Bug*
- square kilometres in Bermuda
- suites in Claridge's Hotel (→191)

Art:

- 'Fifty-three Stations of the Tokaido Highway' – a set of prints by Hiroshige

54

'All-Muggleton had notched some fifty-four.'
Charles Dickens, *The Pickwick Papers*, 1837

'Fifty-four forty or fight' was a slogan in the 1830s, later taken up by the Democrats in the 1844 presidential election campaign, during a boundary dispute between the United States and Canada. According to a treaty of 1818, both countries could occupy the Oregon County, lying between latitudes 42° and 52°40′ north, but in the 1830s and '40s expansionists wanted to take the whole area by force if necessary. The issue was eventually settled with a compromise placing the border between the US and Canada along the 49th parallel.

Fifty-four is also the number of:

- articles of clothing bought by the average American female each year
- basketball riots in the US between 1960 and 1972
- cards in a pack (including jokers)
- countries in which English is an official language
- inches width of a standard double bed
- percentage of new year's resolutions broken within a fortnight
- percentage of the world's buffaloes in India
- sections of the Pentateuch known as Sedra (or Sedrah) one of which is read in the Synagogue at the Sabbath morning service

55

'Fifty-five must seem as old as the hills to a young girl like you.'
Angus Wilson, *Anglo-Saxon Attitudes*, 1956

The fifty-five gallon oil-drum has had a profound influence on the development of civilisation. Quite apart from its primary duty in fuelling the allied forces in World War II, the container itself was then used for a variety of purposes. It was the basic building block for a large number of bridges; it enabled the construction of solar-powered showers at many military bases; and, last but by no means least, it was the inspiration behind the development of the West Indian steel band.

Fifty-five is also the number of:

- cases of wife-sale recorded in Britain between 1840 and 1880
- days it took Johann Hurlinger to walk from Vienna to Paris on his hands in 1900

- inches in the height of Charles I
- murders in Belgrade in 1996 (→40)
- ounces weight of the average human liver
- pints of Carling Black Label drunk every second the pubs are open in Britain
- years' marriage for an emerald anniversary

Film:
- *Fifty-Five Days at Peking* (1963): Charlton Heston and David Niven in the Boxer rebellion

Books:
- *Flying Fifty-Five* by Edgar Wallace (1922)
- *Fifty-Five Years at Oxford*, by G. B. Grundy (1945)

56

'The cat goes with young fifty-six days.'
Oliver Goldsmith, *Animated Nature*, 1774

According to modern veterinary science, Goldsmith was a week out, since the gestation period of a domestic cat is sixty-three days, not fifty-six. Fifty-six is, however, a number that could be of interest to anyone investigating the derivation of the phrase 'cold enough to freeze the balls off a brass monkey'. Although nobody can be certain, one theory is that the balls in question are cannonballs and 'monkey' was a slang term for the stand on which the cannonballs were piled. In conditions of extreme cold, any stand made of brass would contract more than the iron of the cannonballs, thus causing the balls to fall off. In other words, freezing the balls off the brass monkey. Unfortunately, there is a dearth of examples of brass cannonball stands to support this theory. And where does the fifty-six come in? Well, if you make a perfect pyramid of cannonballs, there will be one at the top, supported by three on the level below, and six on the level below that, then ten, fifteen, twenty-one and so on. So a six-storey cannonball pile will have a total of 1+3+6+10+15+21 = 56 cannonballs.

Fifty-six is also the number of:
• days it takes a plucked hair to reappear
• journalists killed in action around the world in 1993
• kilometres the average pencil can write
• minutes in the maximum length of a polo game
• times Salman Rushdie is said to have changed his address in the six months after the fatwa
• years' reign of Henry III

57

'The lunches of fifty-seven years had caused his chest to slip down into the mezzanine floor.'
P. G. Wodehouse, *The Heart of a Goof*, 1926

Fifty-seven is a number that crops up more often than it ought in great disasters. There were fifty-seven people killed by Woo Bum Kong on 28 April 1982 in Sang-Namdo, South Korea, in the world's worst massacre by a single killer; there were fifty-seven people killed in the eruption of Mount St Helens volcano in the United States in 1980 after it had been inactive for the previous 123 years. Fifty-seven was also the number of buckets taken on the Victorian Exploring Expedition from Melbourne, Australia, in 1860 by the team led by Robert O'Hara Burke. They also took twenty-five camels, eighty pairs of shoes, twenty camp beds and thirty cabbage tree hats. The expedition was a disaster.

More auspicious fifty-sevens include the number of:

• books on dentistry published in the UK in 1996
• grams of flour eaten by the average Briton each week
• grams of salt eaten each day by the average horse
• Heinz varieties, allegedly – though there were already over sixty when the 'Heinz 57' catch-phrase was adopted
• police officers in the siege at Glenrowan, Australia, in 1880, that led to the capture of Ned Kelly
• women serving jail sentences in England and Wales for burglary at the beginning of 1997
• world billiard records held by Walter Lindrum when he retired undefeated in 1950

58

'He pretended that he had cleaned up all the tough guys on Fifty-eighth Street.'
J. T. Farrell, *Young Lonigan*, 1936

Fifty-eight is a number associated with extremes of temperature. For –58°C is the average temp of Polus Nedostupnosti in Antarctica – the coldest place on earth – and +58°C is the highest temperature ever recorded, at Al Aziziyah, Libya in 1922.

Fifty-eight is also the number of:

• different languages taught in India's schools
• facets in a 'brilliant cut' diamond
• megatons of a hydrogen bomb exploded by the USSR in 1961, the largest ever built
• novels written by Rider Haggard
• ships sunk by the 1867 hurricane in the West Indies
• tourists murdered by terrorists in Luxor, Egypt, on 17 November 1997

59

'Good engine this. We're doing fifty-nine or an unripe sixty.'
Compton Mackenzie, *The Early Life of Sylvia Scarlett*, 1919

Fifty-nine is a number with good royal connections, for it is the number of kings and queens of Sweden, and also the number of years George III of England reigned.

Fifty-nine is also the number of:

- cities in Britain
- countries in the world that drive on the left
- days in the rotation period of the planet Mercury
- minutes from the opening bell to the end of a fifteen-round boxing match
- percentage of nitrogen in a fart
- square miles of the British Virgin islands
- tons of Berlin Wall shipped to the US in the year following its demolition

Book:

- *Fifty-Nine Icosahedra* (1938) by H. S. M. Coxeter *et al*

60

'Here I sit, alone and sixty,
Bald and fat and full of sin,
Cold I sit and loud the cistern,
As I read the Harpic tin.'
Alan Bennett, *Place Names of China*

The ancient Sumerians and ancient Chinese both did much of their counting in a system to base sixty. The Chinese calendar, in particular, operates according to cycles of sixty years, beginning in 2637 BC, when the legendary Emperor Huangdi invented it. So the seventy-eighth cycle began in 1983 (2637+1983 = 4620, which is 60×77). The sixty seconds in a minute, sixty minutes in an hour and 360 degrees in a circle are all relics of ancient base-sixty counting systems. Plutarch was impressed by the number sixty and believed that crocodiles laid sixty eggs, which incubated for sixty days, before hatching into baby crocs that lived for sixty years.

Sixty is also the number of:

- degrees in each angle of an equilateral triangle
- floral arrangements sent to Gracelands each anniversary of Presley's death
- wives of Khaled Ashta of Egypt. He divorced the sixtieth in 1995 and was reported to be looking for a sixty-first. The marriages lasted between forty-eight hours and three years
- years' marriage for a diamond anniversary

61

'She has thrown her husband out of the house sixty-one times, but he always returned. It looks as if she put too much top-spin on him.'
Punch, 1934

61 Cygni is the 'flying star', the first star other than the sun whose distance from the earth was measured with any accuracy (by Friedrich Bessel in 1838). It is also interesting to note that one person in every sixty-one in Britain is called 'Smith'.

Sixty-one is also the number of:

- centimetres above the ground Louis Breguet's helicopter rose on the first manned helicopter flight in 1907

- centimetres per hour for the speed of a snail
- percentage of Italian men who prefer beer to wine
- points needed to win a game of 8-ball pool
- square miles in Liechtenstein

62

'To reassume an office at sixty-two is not the same as to assume it at thirty-two.'
Matthew Arnold, 1885

Sixty-two is an important number in the history of paying workers piece-rates. An example of this dating back to the Middle Ages was the practice of paying scribes by the 'pecia' (piece), each being precisely sixteen columns, each of sixty-two lines with thirty-two characters to the line.

Sixty-two is also the number of:

- hours a week the average British mother spends on household tasks
- hours a week the average British person over fifteen spends asleep
- references to 'camel' in the Bible
- self-portraits by Rembrandt
- tombs that have been discovered in the Valley of the Kings at Luxor
- years of age of the oldest recorded horse

Sixty-two is also the opus number the French composer Eric Satie gave to his first published composition.

63

'A lady in the virgin bloom of sixty-three.'
Oliver Goldsmith, *The Citizen of the World*, 1762

Sixty-three, being one less than the sixth power of two, is the number of distinct non-empty sets that can be formed from a collection of six objects. Or, to put it in a less mathematical formulation, it is the number of distinct omelettes you can choose to make from one carton of six eggs. (If you work it out, they are six one-egg omelettes, fifteen with two eggs, twenty with three eggs, fifteen with four eggs, six with five eggs, and one with all six.)

Sixty-three is also the number of:

- aeroplanes of the Royal Flying Corps at the start of World War I
- articles of Magna Carta
- gates to the palace in Bangkok of King Rama I of Thailand (reigned 1782–1809)
- heliports in the United States
- inches of Britain's longest moustache
- people fatally shot in the UK in 1994
- possible arrangements of the dots in the Braille system
- sections (tractates) of the Talmud
- types of wren, mostly living in Asia and America; only one type lives in Europe
- years of Victoria's reign; she acceded to the throne on 20 June 1837 and died on 11 January 1901
- completed years of age of Rosanna Dalla Corta when she gave birth in 1994 – at the time the oldest woman to have a child. Her record lasted only three years, when it was broken by a 66-year-old who had lied about her age when applying for fertility treatment

64

'Will you still need me, will you still feed me,
When I'm sixty-four?'
John Lennon and Paul McCartney, 1967

Sixty-four, as the sixth power of two, is simultaneously a square (of eight) and a cube (of four). Its capacity to be halved so many times makes it a natural number to crop up whenever symmetry or divisibility is required. The sixty-four squares of a chessboard and the sixty-four hexagrams of the *I Ching* are both examples of the inherently playful nature of the number, though it was the Indians who took its playfulness most seriously with the sixty-four arts of loving taught in the *Kama Sutra*.

Sixty-four is also the number of:

- calories burnt off by one minute's kissing
- Germans expelled from Great Britain between 1907 and 1914 for procuring and prostitution
- grams of tinned soup consumed by the average Briton each week
- inches height of James Madison, the shortest US President
- kilometres per hour top speed of a kangaroo or an ostrich
- lifts on the London underground system
- squares on a chessboard
- titles in grand slam tennis tournaments won by Margaret Court (twenty-four singles, twenty-one women's doubles, nineteen mixed doubles)

65

'As a rule, the male is generative up to the age of sixty-five, and to the age of forty-five the female is capable of conception.'
Aristotle, *History of Animals*, fourth century BC

Sixty-five is a number with good connections to clothing. It is the number of costumes worn by Elizabeth Taylor in the filming of *Cleopatra*, and also the maximum number of minks needed to make one average-sized coat.

Sixty-five is also the smallest number that can be expressed in two different ways as the sum of two squares of distinct integers: $65 = 1^2+8^2 = 4^2+7^2$ (→50 for the related fact about non-distinct squares).

Sixty-five is also the number of:
- hairs shed daily by the average person
- inches in the height of the average British fully grown male in 1900
- slaves owned by Patrick Henry in 1775 when he said: 'Give me liberty or give me death.'

66

'Sixty-six years ago a vast number both of travellers and stay-at-homes were in this condition.'
Charles Dickens, *Barnaby Rudge*, 1841

Sixty-six is an old German card game, mentioned in the 1857 edition of *Hoyle's Games* as 'Sechs und Sechzig'. By the end of the nineteenth century it had established itself in England. In Islamic counties, however, the number is more highly respected as it is the numerological value (in one ancient system at least) of the name of Allah.

Sixty-six is also the number of:
• Books in the King James Bible
• clickety-click in bingo
• combat missions flown in Korea in 1952 by Edwin 'Buzz' Aldrin
• football riots in the US between 1960 and 1972
• grams of oranges eaten by the average Briton each week
• 'Nylon 66' the name given to the original type of nylon, so-called because both chemicals used in making it had six carbon atoms
• symphonies of the Austrian composer Franz Anton Hoffmeister (1742–1815)
• vehicles per kilometre of road in the UK

Song:
• 'Get Your Kicks on Route 66' by Bob Troup, 1946

67

'You've sixty-seven and you don't cake.'
Rudyard Kipling, *They*, 1904. (The sixty-seven in the quotation were sixty-seven bullocks, and 'cake' means 'to feed on cake'.)

Sixty-seven is the number of African elephants needed to equal the weight of a Boeing 747 Jumbo jet.

Sixty-seven is also the number of:
• knock-out wins by Jimmy Wilde, the 'Mighty Atom', a record for British boxing
• lines in Shakespeare's poem 'The Phoenix and the Turtle'
• novels by Agatha Christie (plus sixteen books of short stories and sixteen plays (→83)
• operatic roles sung by Enrico Caruso
• places called Cripple Creek in the US

• Spanish Armada ships that returned to Spain of the 130 that set out
• US Air Force personnel court-martialled for adultery in 1996 (sixty male and seven female)

68

'It was nothing but a watering depot in the midst of a stretch of sixty-eight miles.'
Mark Twain, *Roughing It*, 1872

Sixty-eight has sporting connections, for it is the maximum length in centimetres of a badminton racquet and also the maximum width in metres of a rugby pitch.

Sixty-eight is also the number of:
• boat race wins for Oxford (1829–1999)
• Germans expelled from Great Britain between 1906 and 1914 for 'housebreaking and frequenting'
• years of hiccups in the longest recorded attack

Film:
• *'68* (1987): a Hungarian opens a café in San Francisco

69

'There are nine and sixty ways of constructing tribal lays, And-every-single-one-of-them-is-right!'
Rudyard Kipling, *Ballads and Barrack-Room Ballads*, 1893

Sixty-nine, or *Soixante-neuf*, is defined in the *OED* as 'simultaneous cunnilingus and fellatio', or by P. Perret in *Tableaux Vivants* (1888) as 'this divine variant of pleasure'. According to the citations in the *OED*, we continued referring to it in French until 1973, though five years later we were already joking about it: 'When I first met him, I thought 69 was a bottle of Scotch' (from the *Guardian Weekly*, 1978). The number sixty-nine is also the only number whose square (4,761) and cube (328,509) between them use each of the digits from 0 to 9 once and once only.

Sixty-nine is also the number of:

- airports in Azerbaijan
- decibels of a loud snore
- kilometres per hour top speed at which a shark has been timed
- people killed at Sharpeville, South Africa, in 1960 when police opened fire on anti-apartheid protesters
- square miles of Washington DC

Films include:

- *Locker Sixty-Nine* (1963): Paul Daneman in detective story
- *69 Minutes* (1977): feature film

70

'The true artist will let his wife starve, his children go barefoot, his mother drudge for his living at seventy, sooner than work at anything but his art.'
George Bernard Shaw, *Man and Superman*, 1903

Seventy is the biblical 'threescore and ten' years of our allotted lifespan, but that is far from being its only significant appearance in the Old Testament. Indeed the Greek translation of the Old Testament was supposedly carried out by seventy scholars (seventy-two according to some sources), which is why it is referred to as the *Septuagint*. Seventy was also the number of men who accompanied Moses to Mount Sinai, and the number of days Moses was mourned.

Seventy is also the number of:
• centimetres length of an average hair before it falls out naturally
• days it took in total to mummify an ancient Egyptian. The word 'mummy' comes from *Mumiya*, Arabic for an embalmed body
• inches height of the average fully grown British male in 1996
• O-level examinations passed by Francis Thomason of West London. Explaining his success at passing more O-levels than anyone else, Dr Thomason stressed the importance of wearing comfortable slippers
• Russians and Poles expelled from Great Britain between 1906 and 1914 for crimes against the person
• William Byrd's compositions for harpsichord included in the 'Fitzwilliam Virginal Book' (*c.* 1625)
• years' marriage for a platinum anniversary

71

The cube of seventy-one is easy to remember. Just write down the odd numbers from 3 to 11 one after another: 357,911, and that's it.

Seventy-one is also the number of:
• grams of pork eaten each year by the average Briton
• men burnt at the stake in Seville for sodomy and bestiality between 1567 and 1616
• the most full hours anyone has spent standing on one leg
• times a year the average Londoner makes love

Film:
• *71 Fragments of a Chronology of Chance* (1994): avant-garde Austrian murder mystery

Autobiography:
• *Seventy-one Not Out* by W. Caffyn (1899)

72

'Furthermore, some men and some women produce female offspring and some male, as for instance in the story of Hercules, who among all his two and seventy children is said to have begotten but one girl.'
Aristotle, *History of Animals*, fourth century BC

Seventy-two was a mystical number in the Middle Ages, when it was claimed in some circles that the name of God had seventy-two letters (though others maintained that He had seventy-two names). Part of the evidence for this belief was three consecutive verses of the Book of Exodus, each of which (in the original Hebrew) comprised exactly seventy-two letters. On the other hand, one might suppose the mystical significance of seventy-two was arrived at more simply by multiplying the three of the Trinity by the twenty-four hours in a day.

Seventy-two is also the number of:

- data collectors employed at the Winbledon Lawn Tennis Championships in 1997
- earthquakes around the world in 1996
- executions personally ordered by Robespierre
- heartbeats per minute for an average adult
- holes in most professional golf tournaments
- letters in the Cambodian alphabet
- points to an inch in type-setters' measurements
- years' reign of Louis XIV of France

73

'Seventy-three' or 'seventy-threes' is American slang for 'best regards' or 'goodbye'. This apparently comes from the habit of Morse code operators to use almost random abbreviations for commonly used expressions. So 'seventy-three' was goodbye and 'twenty-two' was kisses.

Seventy-three is also the number of:

- allied supply ships under the protection of the Dover Patrol sunk in World War II out of 125,000 total
- British killed at Battle of Lexington and Concord, 19 April 1775, the opening battle of the American Revolution
- islands in Ryukyu group of Japan
- liberal democracies among the world's 192 sovereign states in 1995

Film:

- *Winchester '73* (1950), with James Stewart

74

'Seventy-four guineas, Henry. Seventy-four bloody lovely guineas. Just wait till we tell Mr van Huyten about this.'
Stan Barstow, *A Kind of Loving*, 1960

Seventy-four is the number of:

- executions in the US in 1997
- inches length of a standard double bed
- minutes of sound a CD is required to be able to hold
- times Clive Lloyd captained the West Indies at cricket

75

'Listen, my children, and you shall hear
Of the midnight ride of Paul Revere
On the eighteenth of April in Seventy-five.'
Henry Wadsworth Longfellow, *Paul Revere's Ride*, 1863

Seventy-five has two connections with altitude: it is the number of kilometres above the earth's surface where space officially starts and it is also the number of armed helicopters in the Indian air force.

Seventy-five is also the number of:
• paces taken a minute when marching slow time
• pounds worth of Christmas presents the average British under-twelve receives from each parent
• provinces of Turkey
• towns in the world called 'Waterloo', many of which were founded by men who had fought in the battle in 1815

Books:
• *The Cornutor of Seventy-Five* (anon, *c.*1750). A cornutor is a cuckold-maker – one who cornutes, or dallies with other men's wives
• *Seventy-five years Old Virginia* by J. H. Clairborne, 1904
• *75 Brooke Street* by P. Fitzgerald, 1867

76

'Lady Biddy Porpoise, a lethargick virgin of seventy-six.'
Samuel Johnson (in *The Idler*, No. 53, 1759)

Seventy-six is the number of years between successive visits of Halley's comet (or about seventy-six years and thirty-seven days to be less imprecise). When it passed close to earth in 1682 it was observed by Flamsteed, Halley and Hevelius. Halley calculated its orbit and correctly identified it with the comet that had appeared in 1607 and 1531. He predicted that it would return in 1757 (when he would have been 101 years old). In fact, it turned up almost two years late, a delay blamed on disturbances caused by other planets. Its next visit in 1835 was on schedule, and its 1910 appearance was calculated to within two days. Early twentieth-century astronomers traced the history of its sightings back to 240 BC, with other recorded appearances in 87 BC, 11 BC, AD 66, AD 141 and then a long break until 989 and 1066, on which occasion it may be seen depicted on the Bayeux Tapestry.

Seventy-six is also the number of:
• boat race victories by Cambridge (1829–1999)
• British Airways flights diverted in 1996–97 because of medical emergencies
• deaths in the film *Rambo* (of which only one is an American)
• inches height of Lincoln, the tallest US President
• men found guilty of rape in England and Wales between 1805 and 1818
• trombones that led the big parade

77

'At seventy-seven, it is time to be in earnest.'
Samuel Johnson, *A Journey to the Western Isles of Scotland*, 1775

Seventy-seven is the number of:
• inches of snow falling in one day in January 1997 on the village of Montague, NY State – the US record
• provinces in the Philippines
• Russians and Poles expelled from Great Britain between 1906 and

1914 for 'Housebreaking and frequenting'
• times a year the average British adult makes love

And look out for:
• Film: 77 *Park Lane* (1931): Gambling and blackmail in Mayfair, based on the play by Walter Hackett
• Book: 77 *Dream Songs* (1964) by John Berryman
• TV series: 77 *Sunset Strip*

78

Seventy-eight is a good age to attain: it was the age at which Lord Palmerston became the oldest former British prime minister to be cited as co-respondent in a divorce case; it is also the greatest number of years to which an Indian elephant is known to have lived.

Seventy-eight is also the number of:
• cards in the full Tarot pack
• chromosomes of a chicken
• days' record for circumnavigating the world by bicycle
• dollars per heard Gross Domestic Product in Mozambique, the world's lowest
• feet in the length of a tennis court
• gogo bars in Bangkok in 1995
• revolutions per minute of an old gramophone record (discontinued in 1958)

79

'Julia: What is Dr Paramore's number in Savile Row?
Charteris: Seventy-nine.'
G. B. Shaw, *Philanderer* in *Plays Unpleasant*, 1898

Seventy-nine is also the number of:
- different names for the dragonfly in US dialects
- episodes of *Star Trek* in the first three seasons 1966–68

Film:
- *79 AD* (1960): Roman gladiator story

80

'A coachman may be on the very amicablest terms with eighty mile o' females, and yet nobody think that he ever means to marry any vun of them.'
Charles Dickens, *Pickwick Papers*, 1837

Eighty is also the number of:
- chains in a mile
- days around the world in Jules Verne's novel
- grams of pickles and sauces eaten by the average Briton each week
- new mattresses ordered by The Pascha brothel in Cologne in preparation for the 1999 EU summit
- percentage of the world's silk from China
- soldiers in a Roman century (the other 20 had administrative roles)
- victories of Baron von Richthofen, the record for World War I
- weight in pounds of the section of blue-whale penis displayed at the Iceland Phallological Museum
- years life expectancy in Japan, the highest of any major nation, though

according to the *CIA World Fact Book* San Marino (81), Macao (82) and Andorra (83) all offer longer prospects

Film:

• *80 Steps to Jonah* (1969): Mickey Rooney in a tale of theft and mistaken identity

81

'It was the afternoon of my eighty-first birthday, and I was in bed with my catamite when Ali announced that the archbishop had come to see me.'
Anthony Burgess, *Earthly Powers*, 1980, opening sentence

Eighty-one, which is nine squared, is the number of squares on a board for playing Shogi, the Japanese form of chess. It is also the number of stable elements (those having atomic numbers 1 to 42, 44–60, 62–83), and the number of passengers on a Boeing 707 who were killed by lightning in Maryland USA on 8 December 1963.

Most significantly, perhaps, it is the number of times the average British adult makes love each year according to a 1999 survey by Durex.

Film:

• *The 81st Blow* (1975): documentary on oppression during World War II

82

Eighty-two kilometres is the distance between West and East, with East further west than West. That paradoxical statement becomes clear when you measure the shortest distance from Alaska – now indisputably part of Western civilisation – to Russia. America, incidentally, bought Alaska from Russia in 1867 for $7,200,000, which is somewhat less than the film *North to Alaska* (with John Wayne and Stewart Grainger) cost to make ninety-three years later. If you want an even shorter trip from the western world to the East, you have only to sail the four kilometres from Little Diomede, an American Island in the Bering Strait, to its Russian neighbour Big Diomede.

Eighty-two is also the number of:

- metres depth at the deepest point of Lake Victoria
- kilograms weight of a caber
- kilometres in the length of Jamaica from north to south
- lighthouses controlled by Trinity House lighthouse authority
- novels by Erle Stanley Gardner in which Perry Mason solves a case
- recorded accidents involving beds in the UK in 1994
- species of yabby, an Australian crayfish
- specimens of penis in the Iceland Phallological Museum, the world's only museum of male genitalia
- temples restored by the Emperor Augustus

83

In 1950 there were eighty-three cities in the world with more than a million people. Rome is believed to have been the first city to reach the million mark, sometime around the second century BC. Angkor (now in Cambodia) and Hangchow (China) may have reached a million in the early Middle Ages, but there were no million-plus cities anywhere at the beginning of the nineteenth century. London and Paris then became the first modern cities to reach that mark. There are now almost 300 cities (→280) with populations over one million, nine of them in the US.

Eighty-three is also the number of:

- books written by Agatha Christie (sixty-seven novels and sixteen

short-story collections)
• days Harry S Truman served as Vice President before Roosevelt died
• primary schools in St Lucia
• recorded different spellings of 'Shakespeare' by his contemporaries
• television sets per square kilometre in the UK

Film:
• *83 Hours 'Til Dawn* (1990): made-for-TV kidnap drama

84

'On this very second of October, he had dismissed James Forster, because that youth had brought him shaving-water at eighty-four degrees Fahrenheit instead of eighty-six.'
Jules Verne, *Around the World in Eighty Days*, 1873

Eighty-four is also the number of:
• centimetres height of women's high hurdles
• days Calbraith P. Rodgers took to make the first flight across the United States in 1911
• earth years in one orbit of Uranus around the sun
• livery companies in London
• miles length of the world's longest traffic jam – in Japan in 1990
• square kilometres in Lisbon

Films include:
• *84 Charing Cross Road* (1986): Anthony Hopkins and Anne Bancroft in the story by Helene Hanff
• *84 Charlie Mopic* (1989): Vietnam war story

85

Eighty-five is the number of sartorial elegance, for it is the number of different ways a gentleman can knot his tie. This was discovered by two Cambridge physicists, Thomas Fink and Yong Mao, and reported in the science journal *Nature* (vol. 398, p.31) in a short paper entitled 'Designing Tie Knots by Random walks' in January 1999. The number 85 emerges from their investigations into the mathematics of tie-knotting, after limiting the knots (for reasons of practicality) to those involving no more than nine operations. When aesthetic considerations involving the balance of the knot are also taken into consideration, however, the number of tie-knots that are both practical and aesthetic drops to thirteen.

Eighty-five may also be associated with immoral behaviour – for it is the number of Germans expelled from Great Britain between 1907 and 1914 for brothel-keeping; it is the number of lashes to which a Tehran bride was sentenced in September 1995 for dancing with men at her wedding; and it is the number of prostitutes, of the 3,103 examined by the Metropolitan Police in 1837–38, who could read without difficulty.

Eighty-five is also the number of films made by Bette Davis, and the grams of cakes and pastries eaten by the average Briton each week.

86

To 'eighty-six', in American slang, is to refuse to serve an unwelcome customer at a bar or restaurant, and thence, more generally, to throw out or throw away. The derivation is unclear. One theory is that it is simply rhyming slang for 'nix' or nothing. Another idea is that it comes from a more general number system prevalent in the catering industry in the 1920s where various numbers were substituted for commonly used phrases and '86' happened to mean 'we've run out of that'. Yet another version is that it originally referred to a particular bar in Greenwich Village which happened to be at number 86.

It is also the atomic number of Radon, and the number of:

• centimetres of rain and other precipitation that falls on an average spot on the earth's surface each year
• deaths at the storming of the Branch Dravidians in Waco 1993
• metres below sea level of Death Valley, California, the lowest point in the Western world
• ski lifts in Colorado
• square miles of Elba
• stories in the first volume of Grimms' *Fairy Tales* (and another seventy in volume two)

87

• is the 'Four score and seven years ago' referred to by Abraham Lincoln at the start of his Gettysburg Address.

It is also the New York police department precinct number in the Ed McBain series and the number of:

• cases against Britain heard by the European Court of Human Rights 1960–69
• hours of music composed by Verdi
• miles diameter of the biggest meteorite craters on earth (in Canada and South Africa)

88

'Eighty-eight . . . gied you . . . E'en monie a plack, and monie a peck, Ye ken yoursels, for little feck.'

Robert Burns, *Elegy on 1788*, 1789. (Feck is what you need if you are not to be considered feckless.)

Eighty-eight must be the only number to have had a town named after it. In the US Presidential election of 1948, the voting figures for one small town in Kentucky showed eighty-eight people voting for Truman and eighty-eight for Dewey. From that moment on, the town has been known as 'Eighty-eight'.

Eighty-eight is also the number of:

- consecutive baseball games won by UCLA Bruins between 1971 and 1974
- constellations in the sky
- earth days in one year of the planet Mercury
- feet per minute to equal one mile per hour
- keys on a pianoforte
- two fat ladies in bingo

89

Eighty-nine is an important number for California, for it is the number of kilometres of beach along the state's Grand Strand, as well as being the number of Californian condors in the world in 1994.

It is also the atomic number of Actinium, and the number of:

- feet in height of the columns at the Temple of Jupiter in Baalbek.
- times a year the average Singapore adult makes love, the highest in Asia according to a survey by Durex

90

Ninety is a sporty number, on both sides of the Atlantic. It is the number of minutes in a (British) football game as well as being the distance in feet between the bases in baseball.

Ninety is also the number of:
• degrees in a right angle
• executions in Saudi Arabia in the first four months of 1995
• pupils per teacher in primary schools in the Central African Republic (1992) – the world's worst ratio
• women in Turkey seduced by Mozart's Don Giovanni
• years of age at which Sarah conceived Isaac

Films include:
• *90 Degrees in the Shade* (1966): love and corruption in Czechoslovakia
• *90 Days* (1986): Canadian comedy romance

Placename:
• 'Ninety Mile Desert', a limestone area in South Australia and Victoria

91

Ninety-one degrees Fahrenheit is the temperature at which the egg of the mallee fowl incubates. The male digs a hole for the egg in winter and makes a compost heap to generate heat. He regulates the temperature by adding or removing sand from the heap to keep it at a constant 91°F.

Ninety-one is also the number of:
• centimetres of rain that fell at Crowhamhurst, Queensland, on 3 February 1893, the highest rainfall in a day recorded in Australia
• days within which the government promises to repay the holder of a treasury bill
• fatalities at the bombing of the King David Hotel in Jerusalem on 22 July 1946
• kilograms weight for the lower limit of the super-heavyweight class in amateur boxing
• people killed by handguns in Switzerland in 1990

92

There are ninety-two naturally occurring chemical elements and about another twenty that have been created in laboratory experiments, but sometimes they have been so unstable that there is argument about whether they existed. The early history of the chemical elements was one of constant miscounts and recounts. The ancients knew twelve of the elements, but did not know they were elements. In the eighteenth century, Lavoisier listed thirty-three elements, but seven of them were not elements at all. A total of seventy-six elements were added to the original twelve between 1557 and 1925, with a further twenty-one discovered, created or argued about since 1939.

Ninety-two is also the number of:

- countries with capital punishment
- miles of open shelving in the Cambridge University Library
- people killed by hailstones in Gopalganj, Bangladesh, 14 April 1986

Films include:

- *92 in the Shade* (1976): fishing rivalry in Florida Keys
- *92 Grosvenor Square* (1985): World War II spy story with David McCallum and Hal Holbrook

Book:

- *Ninety-two Days* by Evelyn Waugh, 1934

93

Ninety-three is the number of minutes it took to assemble the body of a Model T Ford on the production line. When the Model T was first produced in 1908 its price was $850. The introduction of the world's first production line cut the assembly time from 12½ worker-hours to just over 1½ and by 1913 the price had dropped to $500. It was $390 in 1915 and when, by 1925, the price had dropped to $260, motoring was at last within the reach of the average American family. Between 1908 and 1926 the Model T came in one colour: black.

Ninety-three is also the number of:
• countries in which Citibank has offices
• masses written by Giovanni Palestrina
• Test matches in which Sir Garfield Sobers played
• times Allan Border captained Australia in Test matches
• times Italy was found in breach of the European Convention on Human Rights between 1960 and 1996

Novel:
• *Ninety-Three* by Victor Hugo (set in the year 1793)

94

Few people realise that the number of provinces in Bolivia is identical to the number of State Parks in Michigan. That number is ninety-four, which is also the atomic number of plutonium and the number of:

• baseball games won by Babe Ruth when he was a pitcher in the early part of his major-league career
• inches diameter of the Hubble Space Telescope
• the usual page on which items in the British satirical magazine *Private Eye* are allegedly continued
• people on board James Cook's ship the *Endeavour* in 1768

95

On 31 October 1517 Martin Luther nailed his 'Ninety-five Theses' on the door of the Castle church in Wittenberg. They were statements attacking abuses of the Roman Catholic Church, particularly the practice of selling indulgences, and resulted in Luther being excommunicated and declared a heretic by Pope Leo X in 1521.

Ninety-five is also the number of:

• kilometres per hour in the maximum speed of a dragonfly, the world's fastest insect

• sons of peers killed in World War I by the end of 1914

• *Windows 95*, the computer graphical user interface (GUI) launched by Microsoft at the end of 1995

96

'And there were ninety and six pomegranates on a side; and all the pomegranates upon the network were an hundred round about.'

Jeremiah 52:23

Ninety-six is also the number of:

• different conversations that can be carried on two pairs of telephone wires thanks to the process of 'carrier transmission'

• eggs eaten each year by the average Briton

• height in miles of the angel whose teachings are recorded in the heretical Book of Elkesai

• metropolitan departments of France (plus six more non-European overseas departments)

• minutes Sputnik 1, the first artificial satellite, took to orbit the earth

• storeys of Petronas Towers, Kuala Lumpur

• Test matches played by Rodney Marsh

• victories of Napoleon celebrated on the inner walls of the Arc de Triomphe

Book:

• *My First Ninety-six Years* (1981), an autobiography of the British politician Emanuel Shinwell

97

'Ninety-seven sixpenn'orths of gin-and-water.'
Charles Dickens, *Sketches by Boz*, 1836–37

Ninety-seven is the number of:

• baseball riots in the US from 1960 to 1972
• cigarettes smoked per week by the average British female smoker
• minutes it takes the average person walking briskly to use up the number of calories in a malted milk shake
• passengers on the Hindenburg airship when it exploded on 6 May 1937, of whom thirty-five died
• pounds weight of the weakling in Charles Atlas advertisements who gets sand kicked in his face
• volcanic islands in the Bonin Islands group southeast of Japan

98

Ninety-eight is a significant number in the history of ballooning. In 1937, Jean Piccard, who with his twin brother Auguste was one of the pioneers of high altitude flying in balloons, made the first manned ascent with multiple balloons. He went up in a 'gondola' (an airtight passenger compartment) lifted by ninety-eight balloons each 1.5 metres in diameter.

Ninety-eight is also the number of:

• countries that showed *Dallas* on TV

people per square kilometre in Gambia, Guatemala and Malawi

• teeth of the priodont armadillo, according to Owen in 1854

• tiles in Scrabble with letters on them, plus two blanks

Book:

• *The Trail of '98: A Northland Romance* (1910) by Robert William Service

99

'Genius is one per cent inspiration and 99 per cent perspiration.'
Thomas Alva Edison

Ninety-nine is what doctors ask you to say when they want you to vocalise with an open throat, and what ice-cream van salesmen ask you to say if you want a chocolate flake stuck into your cornet.

Ninety-nine is also the number of:

• centimetres of rain that falls on average in a year in the Bahamas

• days Frederick III reigned in Germany in 1888

• days it took Sir Vivian Fuchs to become the first man to cross Antarctica in 1957–58

• Most Beautiful Names of God, other than Allah, for Muslims

• members of the New Zealand parliament

• the 'Ninety-Nines', an organisation of women pilots founded by Amelia Earhart in 1929, which still exists

• warships commanded by Germany in World War I

• years of age at which the prophet Abraham was circumcised (*'And Abraham was ninety years old and nine when he was circumcised in the flesh of his foreskin'* Genesis 17:24)

• years of the lease Britain had on Hong Kong that ran out in June, 1997

Ninety-nine is also a popular number in film titles, including:

- *The Ninety and Nine* (1922): silent melodrama
- *99 River Street* (1953): boxer struggles to clear his name of murder
- *99 Women* (1969): lesbianism and brutality in a women's jail
- *99 and 44/100 Percent Dead* (1974): black comedy in gangland

100

'There is not one in a hundred of either sex who is not taken in when they marry.'
Jane Austen, *Mansfield Park*, 1814

One hundred, being such a fundamental number in our counting system, is chosen, more often than not, as the number of small currency units in one large currency unit. The following table tells the tale. The first column gives the small unit of currency; the second column is what 100 of these amount to, and the third column is the country of which these are the currency units. (For those countries that have 1,000 small units in a large one, →1,000.)

There are 100:	in a:	in:
cents	dollar	USA, Caribbean, Belize, Brunei, Canada, New Zealand, Fiji, Guyana, Australia, Liberia, Solomon Islands, Zimbabwe
puls	afghani	Afghanistan
centimes	dinar	Algeria
qintars	lek	Albania
lwei	kwanza	Angola
centavos	peso	Argentina, Bolivia, Cuba, Dominican Republic, Guinea-Bissau, Mexico, Philippines
luma	dram	Armenia

groschen	schilling	Austria
gopik	manat	Azerbaijan
kopeks	rouble	Belarus, Russia
chetrums	ngultrum	Bhutan
thebe	pula	Botswana
centimes	franc	France, Belgium, Benin, Burundi, Cameroon, Djibouti, Guinea, Madagascar, Rwanda, Switzerland
fils	dinar	Bahrain
poiska	taka	Bangladesh
centavos	real	Brazil
stotinki	lev	Bulgaria
pyas	kyat	Burma
sen	riel	Cambodia
centavos	escudo	Cape Verde, Portugal
jiao	yüan	China
céntimos	colón	Costa Rica
lipa	kuna	Croatia
cents	pound	Cyprus
haleru	koruna	Czech Republic
øre	krone	Denmark, Norway
dirhams	riyal	Dubai, Qatar
centavos	sucre	Ecuador
piastres	pound	Egypt, Lebanon
centavos	colón	El Salvador
senti	kroon	Estonia
cents	birr	Ethiopia
penniä	markka	Finland
butut	dalasi	Gambia
pfennigs	mark	Germany
tetri	lara	Georgia
pesewas	cedi	Ghana
lepta	drachma	Greece

centavos	quetzal	Guatemala
centimes	gourde	Haiti
centavos	lempira	Honduras
filler	forint	Hungary
aurar	króna	Iceland
paisa	rupee	India, Pakistan
sen	rupiah	Indonesia
pence	punt	Irish Republic
agorot	shekel	Israel
cents	shilling	Kenya
att	kip	Laos
santims	lat	Latvia
lisente	loti	Lesotho
avos	pataca	Macao
paras	dinar	Macedonia
tambala	kwacha	Malawi
sen	ringgit	Malaysia
laari	rufiyaa	Maldives
cents	lira	Malta
cents	rupee	Mauritius
mongos	tugrik	Mongolia
centimes	dirham	Morocco
centavos	metical	Mozambique
pyas	kyat	Myanmar
paisa	rupee	Nepal
cents	gulden	Netherlands, Surinam
centavos	córdoba	Nicaragua
cobo	naira	Nigeria
jun/chon	won	North/South Korea
centésimos	balboa	Panama
toea	kina	Papua New Guinea
céntimos	guarani	Paraguay
cents	new sol	Peru

groszy	zloty	Poland
bani	leu	Rumania
centavos	dobra	São Tomé-Principe
hallala	rial	Saudi Arabia
cents	rupee	Seychelles, Sri Lanka
cents	leone	Sierra Leone
halierov	koruna	Slovakia
stotin	tolar	Slovenia
cents	shilling	Somalia, Tanzania, Uganda
cents	rand	South Africa
céntimos	peseta	Spain
cents	lilangeni	Swaziland
öre	krona	Sweden
piastres	pound	Syria
tanga	rouble	Tajikistan
satang	baht	Thailand
seniti	pa'anga	Tonga
kurus	lira	Turkey
pence	pound	UK
centésimos	peso	Uruguay
tiyin	sum	Uzbekistan
centimes	vatu	Vanuatu
céntimos	bolívar	Venezuela
xu	dong	Vietnam
sene	tala	Western Samoa
fils	rial	Yemen
makuta	zaire	Zaire
ngwee	kwacha	Zambia

One hundred is also the number of:

- pounds of oxygen in the average human body
- the answer if you add together the positions in the alphabet of the letters in 'numeracy'

• women seduced in France by Mozart's Don Giovanni
• yards in an American football field
• years Sleeping Beauty slept

Films include:
• *Anne One Hundred* (1933): romance based on a play by Sewell Collins
• *100 Men and a Girl* (1937): musical
• *The Hundred-pound Window* (1943): Richard Attenborough in a crime drama
• *100 Cries of Terror* (1964): horror story based on works of Edgar Allen Poe
• *100 Rifles* (1969): violent western with Burt Reynolds and Racquel Welch
• *Mama Turns a Hundred* (1979): Spanish comedy with Geraldine Chaplin
• *100% Bonded* (1987): spy film with Sean Connery

101

• Dodie Smith's Dalmatians
• members of the Estonian parliament
• metres length of one side of the square base of the Eiffel Tower
• people per square mile on earth, excluding Antarctica
• quatrains in Edward Fitzgerald's translation of the *Rubaiyat* of Omar Khayyam

Films include:
• *101 Dalmatians* (1961): Disney feature-length cartoon
• *101 Problems of Hercules* (1966): cartoon feature
• *101 Dalmatians* (1996): with Glenn Close as Cruella de Vil

102

is the number of films reviewed in *Cluck – The True Story of Chickens in The Cinema* (1981), by Jon-Stephen Fink, the definitive work on films in which chickens (or their eggs) make an appearance.

102 is also the number of:

- days the average sixty-year-old male has spent shaving
- days the longest recorded case of constipation lasted
- floors in Empire State Building
- verses in the final section of the prose Edda of Snorri Sturluson, each verse illustrating a different metre or stanza form

Film:

- *102 Boulevard Haussmann* (1991): love story based on the life of Marcel Proust

103

The number 103 is connected to one of the more deeply inconsequential numerological coincidences ever discovered. Among the buses that skirt the fringes of Greater London, the 103 goes from Rainham War Memorial to Romford. Now if you add up the positions in the alphabet of the letters in 'Rainham War Memorial', you get 18+1+9+14+8+1+13+23+1+18+13+5+13+15+18+9+1+12 = 192, and if you do the same for Romford you get 18+15+13+6+15+18+4 = 89. Subtract one from the other, and you get 103, the number of the bus.

103 is also the number of:

- Archbishops of Canterbury
- dalmatians tattooed on the back of George Reiger of Pennsylvania. 'We were in a hurry and lost count,' explained Mr Reiger, a Disney fan

• forbidden words in the official (politically correct) Scrabble rules in America, not counting plurals and other derivatives
• Germans expelled from Great Britain between 1907 and 1914 for soliciting or importuning
• kilograms weight of the heaviest mountain lion on record
• kilometres-per-hour wind speed required to reach storm force on the Beaufort scale
• 'nonstellar objects' – galaxies, nebulae and star clusters – listed in Charles Messier's catalogue in 1784
• species of crow

104

There were 104 columns (some say 106) in the Temple of Olympian Zeus, the largest temple ever built in Greece, begun around 530 BC and completed around AD 125. Sixteen of the columns remain today, which curiously enough is the same as the number of stones still standing at Stonehenge (→30).

104 is also the number of:
• characters named John in feature films 1983–93
• pilgrims who set sail to America from Plymouth in 1620
• symphonies by Haydn
• the Tupolev Tu-104, the Soviet Union's first twin-jet airliner, first produced in 1955

105

Element 105 is an artificially produced radioactive element about which the Russians and Americans are constantly arguing. The Russians, who claim they produced it first, want to call it nielsbohrium, in honour of the

Danish physicist Niels Bohr, but the Americans, who are convinced otherwise, propose the name hahnium, after the German chemist Otto Hahn.

105 is also the number of:

- bears Davy Crockett claimed to have killed in seven months
- calories in a large banana
- deaths from scorpion bite in Algeria in 1998
- flights made by the Wright brothers
- times Billy Wright played football for England
- times Shakespeare uses the word 'damned'

106

- disorderly houses involved in prosecutions in England and Wales in 1856–57
- polished diamonds cut from the Cullinan – the largest diamond ever discovered, which was found near Pretoria, South Africa, in 1905 and weighed 1,306 carats
- population of Ellesmere Island, off Greenland, the world's tenth largest island
- times Bobby Charlton played football for England

107

- centimetres height of men's high hurdles
- days in the gestation period of a lion or tiger
- essays written by Michel de Montaigne
- minutes maximum duration of a lunar eclipse
- people per square kilometre in Indonesia or Portugal
- square kilometres in Jerusalem

• verses excluded from the Biblical Book of Esther, which appear as Additions to the Book of Esther in the Apocrypha

108

• beads in a Tibetan rosary
• gallons in a butt of ale
• grams of cheese eaten by average Briton per week
• pounds record weight for a marrow
• minutes Yuri Gagarin was airborne on 12 April 1961 when he became the first man to orbit the earth
• women serving jail sentences in England and Wales for robbery at the beginning of 1997

109

• catches taken by Sir Garfield Sobers in Test matches
• centimetres width of the Turin shroud
• marches written by Edwin Franko Goldman (1878–1956)
• times a year the average male has sex, according to a survey of fifteen countries conducted by *Esquire* magazine
• times the diameter of the earth would fit into the diameter of the sun (which means, of course, that the volume of the sun is more than one million times that of the earth)
• permissible two-letter words in Scrabble, from aa to zo
• car-making companies in the world in 1900 (down from over 300 in 1895)

110

• feet length of the longest blue whale ever measured
• heads of state who gathered in Riocentro, Brazil, in June 1992 for the Earth Summit (UN Conference on Environment and Development)
• men sent to prison in London between 1820 and 1824 for 'indecently exposing their persons'
• pounds paid for two pairs of Queen Victoria's silk stockings at auction in 1978
• storeys of both Sears Tower, Chicago and the World Trade Center, New York

111

• articles in the United Nations charter
• F-111 strike aircraft, not to be confused with the FB-111, a bomber version, or the EF-111, which is described as an 'electronic countermeasure' of the same aeroplane
• kilograms of moon rocks brought back on Apollo 17
• kilometres between two meridians (degrees of longitude) at the equator
• metres height of the Aswan High Dam
• record number of beermats flipped and caught

112

• drainage pumps in the New Orleans flood-prevention system
• chemical elements
• feet of the highest recorded wave
• mountains in Java
• national lottery tickets sold every second in Britain

• pounds in a hundredweight
• short stories by Henry James

113

• days in the gestation period of a pig
• kilometres to the horizon for an observer one kilometre above sea level in a plane or up a mountain
• kinds of plum introduced by the American plant breeder Luther Burbank
• nations in the Non-Aligned Movement

114

• centimetres in an ell, an old measure for cloth
• chapters of the Quran
• cigarettes smoked per week by the average British male smoker
• Coca-Colas drunk per head in the UK in 1995
• radios for every hundred people in the UK
• years the Hundred Years War lasted
• years in a million hours

115

• calories in an ounce of Cheddar cheese
• countries that broadcast Miss World in 1997
• daily newspapers in Sweden
• feet length of the ancient Greek trireme
• grams of fleece produced each year by a vicuna, the smallest member of the camel family, with the finest fleece of any wool-bearing animal

• kilometres length of the river Eure in France
• living rooms in Spruce Tree House, a thirteenth-century ruin in Mesa Verde National Park, Colorado, built by American Indian cliff dwellers
• marine species of angelfish
• miles length of Hadrian's Wall
• species of the Acer genus of tree, including sycamore and maple
• species of butterflyfish

116

• hours it would take to play all the music composed by Purcell
• people per square kilometre in Northern Ireland
• poems in the only surviving manuscript of the Latin lyric poet Catullus. His verses tell of his love for Clodia, an aristocratic maiden whom he refers to as 'Lesbia'. But his passion ended in disillusionment
• square kilometres in Jersey
• square kilometres in Manchester
• state parks in Pennsylvania

117

'Seventy minutes had passed before Mr Lloyd George arrived at his proper theme. He spoke for a hundred and seventeen minutes, in which period he was detected only once in the use of an argument.'
Arnold Bennett

117 was the number of people who founded the 'Lost Colony' at Roanoke Island, off the coast of what is now North Carolina, in 1587. It was England's second colony in America, the first, two years earlier also at Roanoke, having ended in the colonists giving up and returning to England. The 117 of the new colony became 118 on 18 August 1587,

when Virginia Dare was born – the first English person born in America. John White, the leader of the new colony, went back to England for fresh supplies, but the war between England and Spain prevented him from returning to America until 1590. When he arrived, the only traces of the colonists were the letters CRO carved on one tree and the word 'Croatoan' on another. Nobody knows what happened to the 117 lost colonists.

117 is also the number of:

• dual-flush toilets in the Millennium Dome at Greenwich
• the F-117 stealth fighter of the US Air Force
• first-class centuries scored by the Australian cricketer Don Bradman
• German scientists, including Wernher von Braun, sent to the United States at the end of the World War II to work on guided missile systems
• grams of baked beans eaten by the average Briton each week
• kilometres per hour wind speed to be exceeded for hurricane force on the Beaufort Scale
• nations that approved the Law of the Sea Treaty in 1982
• 'You know's counted in a forty-five-minute radio broadcast in Los Angeles in 1996 by Barney Oldfield, a retired Air Force colonel

118

• factories in Turkey in 1923
• inches length of the Willamette meteorite, found in the Willamette Valley, Oregon – the largest meteorite ever found in the United States
• people per sq km in Albania and Armenia
• pieces into which the 'Winged Victory' statue of Nike, the goddess of victory, was broken when it was found in 1863. It dates back to about 180 BC and is now in the Louvre.
• Test matches played by Graham Gooch

119

- airports in Algeria
- caps for Northern Ireland's football team won by Pat Jennings
- kilometres per hour at which a cyclone is called a typhoon
- square kilometres in Dublin
- square kilometres in Ottawa

120

- Days of Sodom in the book by the Marquis de Sade
- hours it would take to play all the music composed by Beethoven
- kilometres the furthest you can be from the sea in Great Britain
- maximum number of members of the College of Cardinals, the body which elects the Pope
- members of the Knesset, the Israeli parliament
- members of the National Assembly of Senegal

121

- countries in which Scrabble is sold
- horse-race winners ridden by Fred Winter in the 1952–53 season
- Methodist preachers in the UK in 1770
- points to win at cribbage
- sexual partners of ancient kings of China: one queen, three consorts, nine wives of second rank, twenty-seven wives of third rank, eighty-one concubines

122

- consecutive victories by Ed Moses in 400m hurdles from 1977 to 1987
- different type-faces designed by Frederic William Goudy (1865–1947)
- metres height of the Lighthouse of Alexandria, one of the Seven Wonders of the Ancient World
- people killed in the floods caused by Hurricane Agnes in 1972
- persons on board the *Dunbar* when she was wrecked on the rocks at Sydney harbour in 1859. Only one survived
- square kilometres of St Helena

123

- fables by the third-century Roman author Babrius
- journalists in jail for their professional activities at the beginning of 1993
- years to which Aaron lived in the Old Testament

Two films have this number in the title:

- *The Taking of Pelham 123* (1974): hostage drama with Walter Matthau and Robert Shaw based on the novel by John Godey
- *123 Monster Express* (1977): murders and a bomb on a bus in Thailand

124

- Japanese emperors in an unbroken line from the same family
- kilometres length of the coastline of Benin
- pounds record weight for a pumpkin
- people per square mile in New Hampshire

125

• baryton trios composed by Haydn (his patron Prince Nicholas of Esterhazy played the baryton)
• calories in a large raw apple
• miles per hour cruising speed of an Intercity-125 train
• plays written by Dion Boucicault (1820–90), an Irish-American actor-manager
• Test matches played by Sunil Gavaskar
• years, from 1835–1960, during which Rio de Janeiro was Brazil's capital

Film:
• *125 Rooms of Comfort* (1974): horror and madness

126

'This note doth tell me of ten thousand French
That in the field lie slain; of princes in this number,
And nobles bearing banners, there lie dead
One hundred twenty-six.'
William Shakespeare, *Henry V*, 1599

126 is the number of:
• centuries scored by W. G. Grace in first-class cricket
• characters named Jack in feature films 1983–93
• gallons in a butt of wine
• people per square kilometre in Andorra and Tonga
• score for Quartzy played as an opening move in Scrabble

127

- hottest temperature on the moon in degrees Celsius
- LZ-127 Graf Zeppelin, the most successful rigid airship ever built
- recorded ways of spelling the surname Raleigh; Sir Walter Raleigh himself is known to have spelt his name as Rawleyghe, Rawley and Ralegh, while his colleagues and friends used Ralo, Ralle, Raulie, Rawlegh, Rawlighe, Rawlye and more than sixty other variants. The one they never seem to have used is Raleigh, which became the dominant spelling through other people of the same name.
- years to which Sarah, Abraham's wife, lived in the Old Testament

128

- Americans killed when the *Lusitania* was sunk in 1915 (→1,198)
- crude gestures in the film *South Park* (1999)
- cubic feet in a cord – a unit for measuring firewood
- metres length of Count Ferdinand von Zeppelin's first airship
- members in the Mexican senate or Lebanon's National Assembly

129

- kilometres per hour top speed of an osprey
- metres height of Montmartre, the tallest hill in Paris
- mystery novels by Erle Stanley Gardner (→82)
- nominations for the 1997 Nobel Peace Prize
- people per square mile in Belarus

130

'If Mr Perry can tell me how to convey a wife and five children a distance of a hundred and thirty miles with no greater expense or inconvenience than a distance of forty, I would be as willing to prefer Cromer to Southend as he could himself.'
Jane Austen, *Emma*, 1816

130 is the number of:

- different names for the oak tree in American dialects
- lynchings in the US in 1901
- test-tube babies born in Australia in 1984 (first year of such)
- words in *Chambers Dictionary* ending in 'lessness'

131

- national scout organisations within the World Organisation of the Scout Movement
- people killed in the 1964 earthquake that hit Alaska
- skyscrapers in New York

132

- Germans captured by Sgt Alvin York at Battle of the Argonne, 8 October 1918
- islands of Hawaii
- rooms in the White House

133

- common councilmen of the Corporation of the City of London

• highest number sharing the jackpot in the UK Lottery
• Indian tribes recognised in Mexico in 1914
• poems in Tennyson's *In Memoriam* (1850)

134

• countries represented at the World Health Organisation conference in Alma-Ata, Kazakhstan, in 1978, which declared the goal of 'Health for All' by the year 2000
• hours of music composed by Schubert
• metres above sea level of the top of the arch of Sydney Harbour Bridge

135

• grams of biscuits eaten by the average Briton each week
• grams of breakfast cereal eaten by the average Briton each week
• times the words 'read my lips' appeared in the *Washington Post* in the first two years of the Bush Presidency

136

'In the spring I have counted one hundred and thirty-six different kinds of weather inside twenty-four hours.'
Mark Twain, Speech to New England Society, 1886

136 is the number of:
• Fahrenheit degrees of the highest temperature ever recorded – in the Libyan desert
• grams of sugar eaten by the average Briton each week
• marches written by John Philip Sousa
• standard tiles in a mah-jongg set

137

• Celsius degrees of the boiling point of acetic anhydride $(CH_3CO_2)O$
• deaths caused by Atlantic hurricanes in 1995
• internal diameter of the dome of St Peter's in Rome in feet
• height in metres of the Old Man of Hoy in the Orkney islands

138

• degrees Celsius to which milk is heated for a few seconds in the process called 'ultrapasteurisation'
• length of the winning world record throw, in feet, by Dewy Bartlett (ex-governor of Oklahoma), in the 1972 world cow-pat throwing championships
• population of Weston-super-Mare in 1801
• references to goat or goats in the Bible
• members of the National Assembly of Madagascar

139

• length of the River Dee, in kilometres, from the Cairngorms to Aberdeen
• years of age of Herbert Badgery, the confidence trickster in Peter Carey's novel *Illywhacker*

140

• beats per minute of a foetal heart
• bottles of wine drunk annually in France per capita
• skeins in a bundle

141

- countries reporting no cases of polio in 1993
- days a new-born grey-headed albatross stays in the nest
- kilometres of the longest Roman aqueduct – in Carthage

142

- books in Livy's *History of Rome*, of which only thirty-five survive
- people killed in the Durban riots of 1949
- pictures by François Lemoyne on the vault of the Salon d'Hercule at Versailles

143

- diameter of the Pantheon in Rome in feet
- grams of fish products eaten by the average Briton each week
- mph of the fastest recorded tennis serve – Greg Rusedski
- square kilometres of Lake Garda, the largest lake in Italy
- volumes Balzac planned for his *La Comédie Humaine* – of which he completed eighty

144

- a gross
- murders in London in 1996
- nuclear warheads under British control in February 1993

145

- kilometres per hour top speed of a bobsleigh
- length of Amiens Cathedral in metres
- Oxford–Cambridge boat races between 1829 and 1999
- area of Lake Como in Italy in square kilometres
- area of Pittsburgh in square kilometres

146

- British prisoners thrown into the 'Black Hole of Calcutta' on 20 June 1756. According to legend, only twenty-three survived, but more reputable research puts the number of dead at only forty-three
- neutrons in the heaviest isotope of uranium

147

- average annual rainfall in Rwanda in centimetres
- horserace winners ridden by Fred Archer in 1874
- maximum break in snooker
- original height of the Great Pyramid of Khufu in metres
- punches landed by Jimmy Carruthers in the 2 minutes 19 seconds it took him to win the world bantamweight championship from Vic Toweel in 1952

148

- length of the coastline of Dominica in kilometres
- members of the House of Representatives in Australia
- Mormon settlers led by Brigham Young to Utah in 1847

• tornadoes in 24 hours in the USA, 3–4 April 1974

149

• degrees Celsius to which milk is heated for UHT (ultra high temperature) pasteurisation
• doctors per 100,000 people in Brazil
• languages spoken in the old USSR
• small islands in the Bay of Islands off the north tip of New Zealand

150

'And the waters prevailed upon the earth an hundred and fifty days.'
Genesis 7:24

150 is the number of:
• days in the gestation period of a goat or sheep
• French expelled from Great Britain between 1907 and 1914 for soliciting or importuning
• mentions of asses in the Bible
• places called Newton in the UK
• psalms in the Bible
• wigs owned by Queen Elizabeth I

151

• the average depth of the Java Sea in feet
• the final psalm, included in the apocrypha of the Greek Orthodox Bible
• length in miles of the Arno river in Italy

• people per square kilometre in Lithuania
• performing artists in the US listed as having communist associations by investigators for Senator McCarthy in 1950
• times a year the average French adult makes love according to a survey by Durex

152

• height of a badminton net at its centre in centimetres
• minutes Neil Armstrong spent on the moon
• references to bullocks in the Bible
• runs scored by W. G. Grace at the Oval in 1880 – the first English Test match century

153

John 21:11 refers to the 153 fishes caught by Christ's disciples, though the significance of the number has never been properly established. One theory holds that if you add the Ten Commandments to the Seven Gifts of the Holy Spirit, you get seventeen, and 153 is the sum of the integers from 1 to 17, but this sounds a little fishy.

153 is also the number of:
• pieces for piano in Bela Bartok's 'Microkosmos'
• area of Minneapolis or the British Virgin Islands in square kilometres
• times round the world all the roads in America would go

154

• days it would take a spacecraft to reach the sun, travelling at a

constant speed of 40,200 kilometres per hour, which is the speed needed to escape the earth's gravity
• the closest distance in kilometres Ernest Shackleton came to the South Pole in his 1908 expedition
• people per square mile in Indiana
• sonnets by William Shakespeare (though there is some doubt about the authorship of the last two)

155

• length in kilometres of the London sewerage system designed by Joseph Bazalgette in 1865
• medical schools in the US and Canada in 1910
• members of the Storting – the Norwegian parliament
• depth in metres below sea level of the lowest point in Africa
• miles per hour that is the maximum speed of a human body in free fall
• miles per second at which the sun revolves around the centre of the galaxy
• newspaper chains in the US
• area of Christmas Island in the Indian Ocean in square kilometres
• area of Staten Island, NY City, in square kilometres
• tribes belonging to the National Congress of American Indians

156

• consecutive weeks for which Ivan Lendl was ranked the world's number one tennis player
• fairy tales by Hans Anderson
• metres length of Westminster Abbey
• Test matches played by Allan Border

157

• height in centimetres of Maureen (Little Mo) Connolly, who became the first woman to complete a Grand Slam in tennis in 1953
• height in centimetres of Napoleon I – about average for a Frenchman at the time, though most high-ranking soldiers and statesmen were taller
• maximum depth in feet of the Sea of Galilee
• height in metres of the spires of Cologne Cathedral
• people per square mile in the Seychelles
• professional tennis singles titles won by Chris Evert

158

• height in metres of Blackpool tower
• length in metres of Milan Cathedral
• unions affiliated to the Australian Council of Trade Unions
• verses in the Greek national anthem

159

'In Perthshire, the idiots are two hundred and eight, the lunatics only one hundred and fifty-nine.'
Sir A. Halliday, 1828

159 is the number of:
• days' holiday a year the ancient Romans enjoyed under the Emperor Claudius
• litres in a barrel of oil
• vessels in the Japanese Navy in February 1993

160

• men needed to move the body of Jumbo the elephant from the tracks in Ontario in September 1885 after he was killed in a railway accident. This was the original Jumbo, after whom all later elephants, aeroplanes and gigantic special offers were named
• perches in an acre
• seconds for the fastest haircut: as performed by Trevor Mitchell of Southampton on 28 October 1996

161

• length of the coastline of Mauritius in kilometres
• new bishops named by the Vatican in 1996
• vessels (including submarines) in the Royal Navy in February 1993

162

• height of the Arc de Triomphe de L'Étoile, in feet
• cells in the brain of the intestinal worm, Ascaris
• length in kilometres of the Suez Canal

163

The Messerschmitt Me 163 rocket-powered plane was developed by Germany in the early years of the World War II. It was capable of speeds of almost 1,000 kilometres per hour and was the prototype for the Me 163 Komet which flew combat missions late in the war.

163 is the number of:
• states with which the Vatican has full diplomatic ties
• villages wiped out when Krakatoa erupted in 1883

164

• poems in *The Temple* (1633) by George Herbert. He also wrote the well-known hymn 'Let all the world in every corner sing'
• references to horses in the Bible

165

• beads in a rosary (15 sets of ten 'Hail Mary's each separated by a Lord's Prayer)
• curtain calls received by Luciano Pavarotti at the Berlin Opera on 24 February 1988
• references to rams in the Bible

166

• members of the Dail Eireann, the Irish parliament
• people injured in train accidents in Britain in the year 1995–96
• references to ox or oxen in the Bible

167

• countries in which Rotary International operates
• height in feet of the planned 'world's tallest Jesus' statue in Troina, Sicily

• letters in the official name of Bangkok: Krung thep mahanakhon bovorn ratanakosin mahintharayutthaya mahadilok pop noparatratchathani burirom udomratchanivetma hasathan amornpiman avatarnsa thit sakkathattiyavisnukarmprasit
• people killed in the Piper Alpha oil rig disaster in 1988
• tennis singles titles won by Martina Navratilova

168

• metres height of the Grand Coulee dam across the Columbia River
• people killed by the Oklahoma City bomb in 1995
• people per square kilometre in Switzerland
• razor blades sold by Gillette in their first year
• square kilometres per television set in Chad

169

• amendments to the constitution of Nevada
• miles of railroad in Luxembourg
• municipalities in Cuba
• towns in Connecticut

170

• metres length of Winchester Cathedral, the longest church in England
• millimetres length of the standard European condom, according to the 1996 European standard EN600 – an increase of about 2 mm on the previous average length
• oarsmen in the crew of a trireme (sixty-two thranite, fifty-four zygiah, fifty-four thalamian oars)

• people who travelled with Henry Ford, and at his expense, to Europe in 1915 to try to persuade the warring nations to seek peace

171

• Gamma Virginis, also called Porrima, in the constellation of Virgo, consists of a pair of fourth-magnitude yellow stars that orbit each other every 171 years.

• 171 is also the number of murders in London in 1995.

172

• countries that competed in the Barcelona '92 Olympics
• miles per hour of the highest wind speed ever recorded in Britain – in the Cairngorms on 20 March 1986
• pounds of potatoes eaten per head each year in the UK
• times the word 'fruit' occurs in the Bible

173

• height in centimetres of a Shire horse
• degrees of frost on the moon in the coldest part of the night
• environmental treaties recorded in the Worldwatch 'Vital Signs 1995' report
• institutes of higher education in Illinois
• metres above sea level of Yding Skovhoj, the highest point in Denmark

174

• British wounded (and seventy-three killed) at the battle of Lexington, the opening battle of the American revolution
• pounds weight of the largest marlin ever caught

175

• age to which Abraham lived in the Bible
• countries represented when Nelson Mandela was inaugurated as South African president on 10 May 1994
• days gestation period of a sheep
• professional fights won by Sugar Ray Robinson (from a total of 202)
• hours music composed by J. S. Bach
• metres depth of the Meteor Crater in Arizona, believed to have been formed about 50,000 years ago by the impact of an iron meteorite weighing 300,000 metric tons
• pairs of legs of the leggiest centipede

176

• earth days for one day on Mercury
• grams of bananas eaten by the average Briton each week
• references to lions in the Bible

177

• days the Roman Games lasted in the middle of the fourth century – ten days of gladiators, sixty-six of chariot races, 101 of theatrical performances

• gold medals won by the UK in the summer Olympics 1896–1996
• kilometres of railway in Borneo
• Russians and Poles expelled from Great Britain between 1906 and 1914 for larceny and receiving

178

• sesame seeds in the average Big Mac bun, according to Ray Croc, founder of McDonald's
• gold medals won by the UK in the Olympics
• vehicles registered to the Russian UN mission in New York

179

• elected members of the Folketing, the Danish parliament
• languages of the Indian sub-continent described in George Grierson's *Linguistic Survey of India* 1903–18
• species of trees counted in a one hectare area of tropical rain forest in South America
• area of Washington DC in square kilometres

180

• degrees in an about-turn
• top score with three darts
• most drinking straws stuffed into a single mouth
• white stones in the set for a game of Go

181

- feet length of the Reclining Buddha in Pegu, Burma
- professional fights won by Henry 'Homicide Hank' Armstrong, the only boxer to hold world titles at three weights simultaneously
- area of Brooklyn in square kilometres
- black stones in the set for a game of Go

182

- ball-boys and ball-girls on duty at Wimbledon 1997
- couples married at Bangrak registry office in Thailand on St Valentine's Day 1995 – the name means 'village of love'
- members of the Belgian senate
- VCs awarded in World War II

183

- degrees below zero Celsius for the boiling point of liquid oxygen
- elected members of the Nationalrat of Austria
- grams of apples eaten by the average Briton each week
- length in miles of the Aara river, the longest in Switzerland
- parishes in Iceland
- shells from the *Kaiser Wilhelm Geschütz* ('Paris gun') that landed within the boundaries of the city of Paris in 1918

184

- Norwegians who voted against independence from Sweden in 1905
- people killed by Hurricane Diana in 1955

• area of Belgrade in square kilometres

185

• defendants in the Nuremberg trials from 1946 to 1949
• length in feet of the Sphinx
• members of the United Nations
• weight in pounds of Sputnik 1, the first ever artificial satellite, launched in 1957
• species of the order Artiodactyla, the even-toed ungulates

186

• diameter in kilometres of Himalia, a small satellite of Jupiter
• record distance in feet that a wellington boot has been thrown

187

• defenders of the Alamo in 1836
• height in feet of the Taj Mahal above its platform
• islands in Franz Josef Land in the Barents Sea – the northernmost part of Russia
• length in kilometres of the River Forth
• length in metres of the Colosseum in Rome
• nominations for the Cleanest Public Toilet award in Singapore in 1997

188

• centimetres of rain that fell in twenty-four hours on Cilaos, on the

island of Réunion in the Indian Ocean during 15–16 March 1952, the highest rainfall ever recorded for a twenty-four hour period
• kilometres into space travelled by Alan Shepard, the first American in space, on 5 May 1961
• passengers, on average, on a non-local train in Britain
• references to lambs in the Bible

189

• carved memorial stones on the Washington Monument
• kilometres of underground tunnels on the Paris Métro
• litres in one barrel of alcohol
• stairs to the top of the pedestal of the Statue of Liberty
• wickets taken in Test matches by Sidney Francis Barnes (1873–1967)

190

• average weight in kilograms of an Indian tiger
• daily newspapers in Argentina
• litres of blood filtered each day by the kidneys
• pony express stations in the US in 1860

191

• bedrooms in Claridge's (→53)
• centimetres maximum permitted cue length in a game of shuffleboard

192

- independent countries in the world in 1995
- metres height of the gateway arch in St Louis
- places called Fairway in the US

193

'There are one hundred and ninety-three living species of monkeys and apes. One hundred and ninety-two of them are covered with hair. The exception is a naked ape self-named Homo sapiens.*'*
Desmond Morris, *The Naked Ape*, 1967

193 is the number of:

- centimetres in height of Charles de Gaulle
- centimetres of snow that fell at Silver Lake, Colorado, in twenty-four hours on 14-15 April 1921, the heaviest snowfall ever recorded
- kilometres in length of Long Island, New York
- lives lost when the *Herald of Free Enterprise* sank off Zeebrugge in 1987

194

- different meanings of 'set' listed in the OED
- films in which Mary Pickford appeared

195

- countries where Coca-Cola is sold
- square kilometres of the Channel Islands

196

• degrees below zero at which nitrogen liquifies
• wingspan in feet of the Boeing 747 Jumbo jet. This is roughly 75 feet longer than the first flight, made by Orville Wright in 1903

197

• centuries scored in his career by Sir Jack Hobbs
• countries competing in the Atlanta Olympics in 1996

198

• people per square mile in Turkey

199

• days in office of President James Garfield in 1881
• executions in the US in 1935

200

• dollars profit made every second by IBM
• grammes of frozen vegetables eaten by the average Briton each week
• mentions of sheep in the Bible
• pounds (or dollars) for passing Go in Monopoly

Film:
• *200 Motels* (1971): Ringo Starr in the life and times of Frank Zappa

201

• length in kilometres of the South Esk River, the longest river in Tasmania
• miles length of the river Clutha, the longest river in New Zealand's South Island
• T1-201, a radio-active isotope of thallium that is useful in diagnosing certain types of heart disease

202

• animal and plant sanctuaries in India
• hours of music composed by Mozart

203

• canons in the 'General Norms' of the Code of Canon Law
• inhabited islands in the Maldives (→1,196)
• members of the House of Representatives in Pennsylvania
• miles above the earth, the maximum reached by Yuri Gagarin in 1961
• miles from the surface of Mars that was the nearest the Mariner spacecraft reached in 1975
• square miles in Bucharest
• square miles in Budapest
• years between James Madison's drafting the 27th Amendment in 1789 and the US Congress ratifying it in 1992

204

• days it took William Willis to sail from Peru to Australia (with a stop in Samoa for repairs) on a steel pontoon raft in 1963–64
• islands in the Andaman group in the Bay of Bengal. The capital, Port Blair, is the only town
• length in kilometres of the longest fjord in Norway. Thanks to all the fjords – narrow inlets of the sea – the total length of the coast is about 21,350 km, which is about half the distance round the world
• weight in metric tons of the Statue of Liberty
• parliamentary penises Lindi StClair claimed to have known in twenty years of business

205

• bones of a horse
• products with Y2K in their names listed among US Patent Office trademark applications in June 1999
• square kilometres of Cincinnati or Osaka

206

• bones in an adult human body, though this number sometimes varies. At birth we each have about 350 bones, some of which fuse together, though the entire process may not be complete until an individual is 25 years old. About one person in twenty has an extra pair of ribs (making 13 pairs in all), and some people are born with only 11 pairs.
• marines guarding the 759 criminals on board the first convict ships to Australia
• mobile phone users per 1000 people in the US

207

• largest number of people ever carried in an airship
• places called Midway in the US
• television broadcast stations in the UK

208

• feet height of the Cape Hatteras lighthouse in North Carolina, the tallest in the US
• idiots in Perthshire in 1828 according to Sir A. Halliday (→159)
• length in kilometres of the River Wye
• people per square kilometre in Malawi
• square miles of Guam in the Western Pacific

209

• stamps stuck on envelopes in five minutes by Dean Gould of Felixstowe to set a new record in 1997

210

• people per doctor in Italy – the world's most doctor-rich country
• width in mm of A4 paper
• '210 Coca-Cola Bottles' by Andy Warhol (sold for $1.9m in 1992)

211

• days Valentin Lebedev remained in space aboard Salyut 7 in 1982–83

• length in metres of the *Great Eastern*, designed by Isambard Kingdom Brunel, which remained the longest ship in the world for 40 years from its launch in 1858

212

• boiling point of water in degrees Fahrenheit. His temperature scale set the freezing point of salt water at zero, and the human blood temperature supposedly at 100, which suggests that he was running a slight temperature that day
• members of the Belgian Chamber of Representatives

213

• feet of the highest point on Anguilla

214

• Batman articles on sale in 1989
• meanings of the word ‘ka’ in Japanese

215

• daily newspapers published in Belarus
• area of Cairo or Newcastle in square kilometres

216

• miles length of the Laramie river
• pounds of sugar eaten each year by the average Israeli (the world's greatest sugar eaters)

218

• grains in the weight of a biblical shekel

219

• millimetres diameter of a discus

220

•people treated each year in New York for rodent bites
• yards in a furlong

221

• passengers killed in crashes on British airlines 1970–79
• scenes of violence in the film *South Park* (1999)
• volcanoes in the Philippines, of which fifty are active

222

• fatalities when the bullring collapsed in Sincelejo, Colombia, on 20

January 1980, the world's worst bullfight disaster
• offences punishable by death in Great Britain in 1819

223

• mountains over 2,000 feet in Vermont

224

• minutes the average Brazilian spends each year making international phone calls

225

• bronze medals won by the UK in the summer Olympics 1896–1996
• earth days in one year on Venus
• grams of fresh green vegetables eaten each week by the average Briton

226

• length in feet of the 'Long Man' of Wilmington, East Sussex

227

• people killed in the UK's worst rail crash, at Quintinshill, Scotland, in 1915
• ways of cooking chicken in Escoffier's *Guide Culinaire* 4th edition (1921)

228

• American slang terms for drunkenness listed by Benjamin Franklin in 1737

229

• minutes television watched by the average Briton every day
• people found guilty of murder in England and Wales between 1805 and 1818
• pounds of meat eaten each year by the average Austrian, Europe's greatest meat-eaters
• Xerox copies made on average for each of the 35,000 delegates to the UN Women's Conference in Peking in 1996

230

• lynchings in the US in 1892, the record for a calendar year
• Olympic swimming gold medals won by the US, 1896–1996

231

'Pacific 231' is a 'symphonic movement' by Arthur Honegger, composed in 1923. Naming his work after an American railway engine, the Swiss composer said: 'I have not aimed to imitate the noise of an engine, but rather to express in terms of music a visual impression and physical enjoyment.'

231 is also the number of:
• miles per hour wind speed recorded at Mount Washington

observatory on 12 April 1934, at the time 'the highest wind ever observed by man' as is recorded on a plaque at the observatory
• years the Lord Chancellor acted as official censor of plays (1737–1968)
• women in Germany seduced by Mozart's Don Giovanni

232

• airports in Peru

233

• silver medals won by the UK in the Summer Olympics 1896–1996

234

• men arrested in Paris in 1749 for homosexuality

235

• isotope of uranium used in making an atom bomb

236

• categories for the unemployed in the records of the US Census Bureau
• miles an hour wind speed recorded in Guam during a typhoon in 1999

237

- grams of poultry eaten each week by the average Briton

238

- people per square kilometre in the UK
- towers in the Turkish city of Iznik
- kilometres of tram route in Melbourne, Australia

239

- sled dogs in harness that pulled a 12.5 ton truck 50 metres on 10 September 1995 to set a new record
- area of Singapore in square miles

240

- collisions in North America each year between aircraft and Canada geese
- hours it would take to listen to all the music composed by Mozart

Film:
- *240-Robert* (1979): plane crash suspense

241

- Buddhist temples in Los Angeles

242

• feet in length of the pinnacles on the Petronas Towers in Kuala Lumpur
• pounds average weight of an Asiatic black bear

243

• days in the rotation period of the planet Venus, which, like Uranus, rotates about its axis in the opposite direction to its rotation about the sun

244

• miles length of the Mason-Dixon line, drawn by Charles Mason and Jeremiah Dixon in 1763 to settle a border dispute between Pennsylvania and Maryland
• trespassers or suicides killed by trains in Britain in the year 1995–96

245

• kilometres length of Lake Reindeer, Canada
• vehicles per kilometre of road in Taiwan

246

'It was the business of the All-British Company to produce seventeen exceptionally bad and cheap films every year in England in order to allow two hundred and forty-six exceptionally bad and expensive films to be imported every

year into England from Hollywood. This is called the Quota System.'
A. G. MacDonnell, *How Like an Angel*, 1934

- 246 is also the number of people killed by hailstones in Moradabad, N. India in a storm in 1880.

247

- wing-flaps per second of a bee

248

- earth years in one orbit of Pluto around the sun
- solar eclipses that will be visible from earth during the 23rd century

249

- miles on the New York subway system

250

- hedgehogs to equal the weight of one lion

251

- miles on the London Underground network
- people killed in the earthquake in Yunnan, China, on 3 February 1996

• towns and cities in Vermont, where 'The 251 Club' tries to encourage people to visit them

252

• acres of Bronx Zoo when it opened in 1899. It now covers 265 acres

253

• people you need to encounter to have a greater than even chance of meeting one who shares your birthday

254

• metres drop of Takhakaw Falls in British Columbia

255

• names in the first phone book produced by the London Telephone Company in 1880

256

• acres in a hide (a twelfth-century land measure)
• cycles per second of middle C in the scientific scale. Concert pitch places middle C just below 262 cycles per second
• tambourines played in 20.47 seconds by Rowdy Blackwell on 28 October 1996 in a record-breaking performance for charity

257

• people killed in the worst ever plane crash in Antarctica in 1979

260

• days in the Aztec religious calendar

261

• bastards born in Rochdale, Lancashire, between 1661 and 1720
• pounds of meat eaten in the US per person each year

265

• people per square mile in Pennsylvania

266

• average number of days from conception to birth of a human being

267

• lowest four-round total in the British Open Golf Championship as scored by Greg Norman in 1993

268

'Two hundred and sixty-eight sequins are more than I dare lay out.'
Horace Walpole, 1759

- 268 is also the number of words in the Gettysburg Address.

270

- record score in the US Masters Golf Championship set by Tiger Woods in 1997

271

- gold medals awarded at the Atlanta 1996 Olympics

273

- diameter in miles of asteroid Hektor, the biggest discovered this century
- seconds of silence in John Cage's composition '4 min 33 sec'
- people per square kilometre in Israel

274

- average walking speed of a woman in feet per minute
- eggs eaten each year by the average person in the Czech Republic
- stations on the London underground

275

- average walking speed of a man in feet per minute

276

- murders in Dallas, Texas in 1995
- breeds of bird in the Faeroe Islands

277

- atomic mass of the element with atomic number 112, the heaviest ever. One atom was created in 1996 and lasted a third of a millisecond

279

- words in the Ten Commandments in German
- convicts on board the last British convict ship to dock in Australia in 1868

280

- cities with more than one million people at the beginning of 1997

282

- members of the Canadian House of Commons

284

- bank robberies in Rio de Janeiro in 1997

285

- participants in the 1896 Olympics

286

- Marks and Spencer stores in the UK
- verses in chapter two of the Quran, its longest chapter

288

- gallons in a chaldron

290

- women serving jail sentences in England and Wales at the start of 1997 for violence against the person

292

- offensive epigrams written by John Davies (1565?–1618), a poet and writing-master of Hereford. His comments on his contemporaries were described as 'scoundrelisms' by F. E. Hulme in 1902.

295

• gold medals competed for at the Sydney 2000 Olympics

297

• length in mm of A4 paper

299

• pairs of size 12 Bruno Magli shoes ever sold – an important piece of evidence at the trial of O. J. Simpson

300

• golden bees found in the tomb of Childeric, an eighth-century king of the Merovingians

Films include:
• *The 300 Spartans* (1962): re-enactment of the Battle of Thermopylae in 480 BC
• *The 300 Year Weekend* (1971): drama

303

• escalators on the London underground before the Jubilee Line Extension boosted the number to 417
• hours it would take to play all the music composed by Handel
• deposits lost by the Natural Law Party in the 1992 UK general election

305

• lightbulbs on the clock above Times Square registering the US national debt
• mentions of 'wilderness' in the Revised Standard Bible

312

'One out of three hundred and twelve Americans is a bore, for instance, and a healthy male adult bore consumes each year one and a half times his own weight in other people's patience.'
John Updike, *Assorted Prose* (1965) 'Confessions of a Wild Bore'

• 312 is also the number of sports-related riots in the US between 1960 and 1972.

315

• bombs dropped on Germany in World War II for every bomb that fell on Britain
• days in the gestation period of a camel

317

• length in kilometres of the average domestic flight by a UK airline (→1,032)

Film:
• *317th Platoon* (1965): French anti-war film about Vietnam

318

• times the mass of the earth to equal that of Jupiter

319

• pencil-makers in Great Britain at the time of the 1851 census

321

• kilometres a year walked by the average Briton

325

• albums in George V's stamp collection
• miles travelled each year by train by the average Briton

327

• words a minute spoken by J. F. Kennedy in a speech in 1961

328

• feet length of the world's longest sushi roll, made in San Francisco in 1999

329

• feet height of a Coast Douglas fir in Coos County, Oregon – the tallest tree in the US

330

• Minke whales killed by Japan in 1990

336

• gestation period of a horse in days

340

• days in the gestation period of a llama
• hours it would take to play all the music composed by Haydn, the most prolific of the major composers
• sheep per person in the Falkland Islands

341

• suicides from the Eiffel Tower in its first 75 years

343

• Coca-Colas drunk per head in the US in 1995

349

- languages the entire Bible has been translated into
- record number of pancake tosses in two minutes

352

- days in the gestation period of a badger

353

- farms in Wales affected by the Chernobyl disaster
- triremes in the Ionian Greek fleet at the battle of Miletus in 499 BC (→170)

357

- pawnbrokers' shops in Thailand

Film:
- *357 Magnum* (1977): murder, robbery and espionage

360

- degrees in a circle
- different plants eaten by Santa's reindeer according to the official Santa in Lapland – but they don't eat carrots

364

• alleged number of illegitimate children of Friedrich August I of Saxony (King Augustus II of Poland)
• total number of presents received in the song 'Twelve Days of Christmas'

365

Best known as the number of days in a non-leap year. The ancient Babylonians and Egyptians knew that the earth took about 365-and-a-quarter days to orbit the sun, but Europe generally muddled along with a 365-day year until Julius Caesar introduced the idea of leap years (an idea he picked up from Cleopatra). By adding an extra day to February every fourth year, the Julian calendar went a long way towards compensating for the fact that the earth's orbit round the sun takes not 365 but 365.2422 days. The Christian Church adopted the Julian calendar at the council of Nicaea in AD 325. The next correction came in 1582 from Pope Gregory's mathematicians and astronomers who calculated the present system of omitting the leap day in century years, except when the new century is divisible by four. (So 1800 and 1900 were not leap years, but 2000 is.) The new system guaranteed 146,097 days every 400 years, which works out at an average 365.2425 days a year. Despite the accuracy of that figure, it took Britain another 170 years to agree to it. And there were still many 'Give us back our eleven days' protests about the dates that had to be dropped to make up for having had the wrong calendar since AD 325.

365 is also the number of:

• days gestation of an ass
• steps leading to the top of St Paul's Cathedral
• years lived by Enoch in Genesis

Film:
- *365 Nights in Hollywood* (1934): comedy thriller

366

- days in a leap year
- height in feet of the dome of St Paul's Cathedral

368

- security guards and commissionaires employed at the Wimbledon Lawn Tennis Championships in 1997

370

- different cheeses in France (cited by de Gaulle as the reason the country could not be governed)
- dollars per person per year in the gross national product of Haiti

371

- cars per thousand people in the UK

375

- asteroids discovered by K. Reinmuth of Heidelberg
- runs scored by Brian Lara for the West Indies v England, Antigua, 16–18 April 1994, the highest individual innings in a Test match

377

• weeks Steffi Graf was number one in the women's world tennis rankings

379

• black spots on Moroccan roads – at each of which at least ten accidents and ten deaths or serious injuries have occurred within five years
• people per square kilometre in the Netherlands, the most densely populated country in Europe

380

• stones on which the Chinese calligrapher Chen Zhaoguo carved the Bible between 1985 and 1995. It needs magnification fifteen times to be legible

394

• height in feet of the Buddha statue in Tokyo

398

• recorded injuries connected with toilet seats in the UK in 1993

399

• times a word ruder than 'damn' is uttered in the film *South Park*

400

• *The 400 Blows* (1959) – François Truffaut's first feature film

404

• number of ways Brightlingsea has been spelt since its Celtic origin of Brictrich

413

• beds owned by Louis XIV

420

• people per doctor in the USA
• most kittens produced by one female cat

427

• record number of two-egg omelettes made in 30 minutes

432

• pints in a hogshead

434

• Test wickets taken by Kapil Dev, the record for any bowler

435

• members of the House of Representatives in the United States
• television sets per 1,000 people in the UK
• vasectomies performed in a six-month period, from March to September 1995 in Iran

440

• cycles per second for concert pitch A above middle C following an international agreement in 1939

442

• airports in Australia

449

• people killed in earthquakes throughout the world in 1996

450

• pet cemeteries in the US

451

• *Fahrenheit 451* by Ray Bradbury – the novel's title comes from the temperature at which paper catches fire
• seconds a solar eclipse may last

452

• asteroids identified in the nineteenth century

453

• area of Andorra in square kilometres

464

• the lowest temperature on Venus in degrees Celsius

466

• height in feet of spire of Strasbourg Cathedral – the tallest structure in France apart from the Eiffel Tower

468

• average annual rainfall in inches at Mawsynram, India, a strong candidate for the title of the wettest place on earth

469

• stations on the New York subway

471

• weeks in the UK charts for 'Bat Out of Hell' by Meatloaf

472

• beaches in the UK

483

• cinemas in Great Britain

485

• babies abandoned by their mothers in public places in Thailand in 1998

487

• cars per thousand people in the US

488

• average number of beans in a standard Heinz tin (→494)

491

• *491* (1963) – Swedish teenage film drama

494

• average number of beans in a Tesco tin (made by Heinz) (→488)

500

• sheets in a ream
• Indianapolis 500

Film:
• *The 500 Pound Jerk* (1972): US Olympic weightlifter falls in love with Russian gymnast

501

• score required to win a leg at darts

503

• pounds spent per visit to Britain by the average tourist
• metres span of the Sydney Harbour Bridge

520

• cells in Pentonville prison when it was built in the 1840s

530

• cubic feet of air used in a day's breathing by one person

533

• patented inventions of Edwin Land, inventor of the Polaroid Camera

555

• feet length of St Paul's Cathedral

563

• feet high statue of Chief Crazy Horse by Korczak Ziolkowski at Thunderhead Mountain, S. Dakota

565

• days a female bedbug can go without food

570

• convictions for rape in the UK in 1996

574

• bottles of claret William Pitt the Younger is said to have drunk in a single year (→854)

575

'Phileas Fogg, having shut the door of his house at half-past eleven, and having put his right foot before his left five hundred and seventy-five times, and his left foot before his right five hundred and seventy-six times, reached the Reform Club.'

Jules Verne, *Around the World in Eighty Days*, 1873

576

• dots on a computer screen needed to make all the Japanese characters (→35)

577

• lynchings in Mississippi 1882–1956
• stairs in the annual 'Supra Ultra Stairs Climbing' event up the Fukuoka Tower in Japan

579

• years it took to build Milan cathedral

582

• miles diameter of Ceres, the largest known asteroid

583

• dollars per capita spent each year on defence in the UK (→1,074)

587

• wing flaps per second of a mosquito

594

• miles walked each year by the average housewife in the course of her duties

600

- ways to make love, according to the Marquis de Sade
- minimum weight of women's javelin in grams

618

- caravans in Freemont, California

619

- area of Singapore in square kilometres

621

faulty 'proofs' of Fermat's Last Theorem submitted in first year of the Wolfskehl Prize in 1908. The prize, for a proof of the conjecture that the equation $x^n+y^n = z^n$ cannot be satisfied for any integers, x, y, z and n, with $n>2$, was offered by a man whose fascination with the problem had saved him from suicide. It was finally won in 1997 by the English mathematician Andrew Wiles.

632

- species officially listed as endangered in the first 20 years of the Endangered Species Act (1973–93)

633

• *633 Squadron* (1964) – film in which twelve RAF Mosquito pilots bomb a Norwegian cliff to destroy a Nazi munitions factory on a fjord

634

• VCs awarded in World War I

635

• square centimetres that could be allocated to each person, if all the world's population were on the Isle of Wight. Which means that it may just be possible for everyone in the world to stand up simultaneously on the Isle of Wight. With each person having to squeeze into the equivalent of a rectangle about the size of an A4 sheet of paper, however, it would be extremely uncomfortable.

639

• named muscles in the human body

640

• acres in a square mile
• oilwell fires in Kuwait at the end of the Gulf War in 1991

641

an important number in mathematics. Pierre Fermat, the greatest of seventeenth-century mathematicians, noticed that two plus one, and two-squared plus one, and two-to-the-fourth plus one, and two-to-the-eighth plus one (3, 5, 17 and 257) were all prime numbers. He conjectured that all numbers of the same form – two raised to a power of two and one added to the answer – were prime. This was disproved by Leonard Euler, who found that $2^{64}+1$ (a number with nineteen digits) is divisible by 641.

642

• official engagements carried out by Princess Anne in 1997

650

• maximum voltage of an electric eel
• miles per second Santa's sleigh has to travel to deliver all the presents on Christmas Eve

659

• seats in the House of Commons

660

• gestation days of an African elephant

665

• odd gloves found on London Transport 1997–98

666

'Here is wisdom. Let him that hath understanding count the number of the beast: for it is the number of a man; and his number is Six hundred threescore and six.'
The Revelation of St John the Divine, 13:18

Perhaps the strangest thing about 666 being the number of the beast is its connection with the other number of Life, the Universe and Everything, 42. The link is made in Tolstoy's *War and Peace* when Pierre reveals a secret revealed to him by 'one of his brother Masons'. A few verses earlier than that quotation from Revelation we read: *'And there was given to him a mouth speaking great things and blasphemies; and power was given to him to continue forty and two months.'* Now if we write out twenty-five letters of the alphabet (our twenty-six with 'i' and 'j' identified), then assign the numbers 1-9 to the letters a-i, then go up in tens from k (=10) to z (=160), when you add up the letters of 'L' Empereur Napoleon', you end up with the number 666. *'Moreover, by applying the same system to the words "quarante-deux", which was the term allowed to the beast that "spoke great things and blasphemies", the same number 666 was obtained; from which it follows that the limit fixed for Napoleon's power had come in the year 1812 when the French emperor was 42.'*

Macaulay, however, refused to accept this, or other equally ingenious, ways of pinning the number 666 on Napoleon. He preferred to associate the House of Commons with the Beast, because it had 658 members, three clerks, a sergeant and his deputy, a chaplain, a doorkeeper and a librarian, making a total of 666.

In 1995, hundreds of Rumanian peasants refused free shares in newly privatised industries because their serial numbers began with 666.

666 is also the average number of rolls of lavatory paper used each day in the Pentagon.

670

• Martian days in a Martian year (more precisely, it is 669.774)

673

• men involved in the Charge of the Light Brigade

675

• pieces of music written by Mozart

677

• homicides in England and Wales in 1994

678

• area of Bahrain in square kilometres

685

'Yet, who can help loving the land that has taught us Six hundred and eighty-five ways to dress eggs?'
Thomas Moore, 1779–1852

• 685 is also the number of people killed in fires in the UK in 1998

687

• earth days in a year on Mars

688

The number of Friday the Thirteenths every 400 years. Because the Gregorian calendar (→365) provides 146,097 days in its 400-year cycle, and 146,097 is divisible by 7, it means that one complete cycle will take exactly 20,871 weeks, and the next cycle will begin on the same day of the week. Since there are 4800 months in 400 years, each of which has one thirteenth day, and 4,800 is not divisible by 7, there must be an imbalance in the distribution of those thirteenth days among the seven days of the week. In fact, there are 685 Monday the Thirteenths, 685 Tuesday the Thirteenths, 687 on Wednesday, 684 on Thursday, 688 on Friday, 684 on Saturday and 687 on Sunday. So the thirteenth of the month is more likely to fall on a Friday than any other day of the week.

696

• bastards born in Rochdale, Lancashire, 1581–1640

710

- pounds record weight of a pumpkin

711

- *711 Ocean Drive* (1950) – film with Edmund O'Brien as a phone-tapper

725

- homicides in England and Wales in 1991

728

- centimetres length of the world record throw of a toilet bowl by Lars Stene at Namsos, Norway, in 1999

732

- cars per thousand people in the Lebanon, the world's highest proportion
- survivors when the *Titanic* sank in 1912

736

- convicts aboard the first British convict ship to land in Australia in 1788

737

• number of a Boeing jet
• cows in Pennard, Glastonbury, Somerset, milked to make a 9 ft diameter cheese for Queen Victoria in 1841

741

• weeks in the US charts for Pink Floyd's 'Dark Side of the Moon'

746

• most words on a postage stamp – a Greek stamp issued in 1954

750

• herrings in a cran (approximately). A cran is the unit of volume for fresh-caught herrings. It is believed to be derived from a Scottish Gaelic word meaning 'portion' or 'allocation'. Under the Cran Measures Act of 1908, fresh herring had to be sold by the cran in certain places in the UK

756

• grams of bread eaten by the average Briton each week

759

• miles travelled each year by the average Briton going shopping

762

• Boeing 747s you would need to carry all of Bill Gates's money if it were changed into $1 bills

763

• people mentioned in the closing credits of 'Who Framed Roger Rabbit?' – but not Kathleen Turner, who was the voice of Jessica Rabbit

777

'. . . the slightest consideration will show that though seven hundred and seventy-seven is a pretty large number, yet when you come to make a teenth of it, you will then see, I say, that the seven hundred and seventy-seventh part of a farthing is a good deal less than seven hundred and seventy-seven gold doubloons.'
Herman Melville, *Moby Dick*, 1851

• 777 is also the number of years to which the Old Testament prophet Lamech lived.

780

• number of ostriches in Australia in 1922 according to government statistics

793

• centimetres length of the longest recorded human hair, on the head of Swami Pandarasanaj in 1949

800

• minimum weight of men's javelin in grams
• warriors in Valhalla

Films include:
• *800 Leagues Down the Amazon* (1993): adaptation of a Jules Verne adventure
• *800 Leagues Over the Amazon* (1960): another Jules Verne adaptation
• *800 Two Lap Runners* (1994): Japanese drama of athletics and love

803

• grams of potatoes eaten by the average Briton each week

812

• three-letter words in current use in English

815

• televisions per 1,000 people in the US in 1992

828

• murders in Los Angeles in 1995

830

• millimetres of rainfall at the average spot in England and Wales in 1995

840

• pedestrians run down and killed in Bogotá, January 1995 to June 1996; drunken walking was made an offence in November 1996

844

• Disney tattoos on George Reiger of Pennsylvania (→103)

850

• fixed stars catalogued by Hipparchus in 129 BC
• number of words in the Basic English proposed by Professor C. K. Ogden in 1930

851

• fifteen-letter words in current use in English

854

• bottles of Madeira William Pitt the Younger is said to have drunk in a single year (→2,410)

855

• special terms and euphemisms used by the Stasi secret police in East Germany

870

• length in feet of the side of a cube big enough to hold all the human blood in the world (according to John Allen Paulos)

871

• catches taken by W. G. Grace

888

• children reputedly fathered by Moulay Ismail (d. 1727), emperor of Morocco

911

• the number to dial for emergencies in the US

915

• aircraft lost by the RAF in the Battle of Britain

928

• fishmongers' shops in Great Britain

930

• years to which Adam lived according to the Old Testament

931

• people personally killed by Behram, leader of the Thugee cult in India 1790–1830

934

• pounds a minute spent on Viagra worldwide in the first quarter of 1999

946

• company directors banned from running limited companies in the UK in 1996

950

• years to which Noah lived in the Old Testament

962

• years to which Jared lived in the Old Testament

963

• length in feet of the *QE2* cruise liner

969

• years to which Methuselah lived – the greatest age of anyone mentioned in the Bible

976

• the dialling code for the underworld in the following two films:
• *976-Evil* (1988): shy teenager contacts Satan on the telephone
• *976-Evil 2: The Astral Factor* (1991): Satan calls back

978

• days of the longest recorded attack of sneezing

981

- women killed in road accidents in the UK in 1994 (→2,251)

989

- *984: Prisoner of the Future* (1984) – post-nuclear-holocaust prison film drama

989

- accidents on British railways in the year 1995–96

1,000

- words a picture is worth
- ships launched by face of Helen of Troy

Some currencies have 1,000 small units in a large unit:

There are 1,000	in a	in
millimes	dinar	Tunisia
millièmes	pound	Sudan
baiza	rial saidi	Oman
dirhams	dinar	Libya
fils	dinar	Jordan, Iraq, Bahrain
escudos	peso	Chile

1,000 is also popular in the cinema:
• *The House of a Thousand Candles* (1936): from the spy story by Meredith Nicholson
• *Night Has a Thousand Eyes* (1948): Edward G. Robinson in a tale of suspense
• *I Died a Thousand Times* (1955): Jack Palance as a gangster with a heart of gold
• *Man of a Thousand Faces* (1957): film biography of Lon Chaney Jr, with James Cagney
• *These Thousand Hills* (1959): western based on the novel by A. B. Guthrie, Jr
• *The Thousand Eyes of Dr Mabuse* (1960): Fritz Lang horror
• *A Thousand Clowns* (1965): comedy that won a Best Supporting Actor Oscar for Martin Balsam
• *Night of a Thousand Cats* (1972): madman breeds flesh-eating cats
• *1,000 Roses* (1994): a Dutch allegory with a town overrun by roses and the hero going mad

1,001

• Arabian nights
• name of a British carpet cleaning fluid, with the pre-decimal currency advertising slogan: 'One Thousand and One cleans a big, big carpet for less than half-a-crown'

Film:
• *A Thousand and One Nights* (1945): Cornel Wilde as Aladdin

1,003

• women in Spain seduced by Mozart's Don Giovanni

1,016

• recorded injuries in the UK in 1993 involving baths

1,032

• kilometres in the average length of a flight on a UK airline (→317)

1,046

• winter population of Antarctica

1,049

• highest score in a game of Scrabble, as recorded by Phil Appleby in July 1989

1,056

• deaths on passenger flights in 1997

1,071

• *1,071 Fifth Ave* (1994): documentary film on the life of Frank Lloyd Wright

1,074

• dollars per capita spent each year on defence in the US (→583)

1,082

• establishments in Bangkok in January 1995 offering sexual services

1,092

• television broadcast stations in the US

1,093

• patents in the name of Thomas Edison

1,110

• fires every day attended by fire brigades in the UK in 1998

1,114

• most votes polled by Screaming Lord Sutch (of the Monster Raving Loony Party) in a British by-election (Rotherham 1994)

1,146

• radios per 1,000 people in the UK in 1992

1,163

• weight in tons of the statue of Christ in Rio de Janeiro

1,173

• prisoners taken by Cuba at the disastrous Bay of Pigs invasion in 1961

1,177

• centimetres in the average annual rainfall at Tutunendo, Colombia, another claimant (→468) for the title of the wettest place on earth

1,182

• murders in New York in 1995

1,189

• chapters in the Bible
• persons prosecuted between December 1719 and November 1720 for 'lewd and disorderly practices' by the Societies for Promoting a Reformation of Manners

1,190

- patients vasectomised in one day in Bangkok in December 1983

1,196

- islands in the Maldives (→203)
- people killed on roads of Morocco in first six months of 1996

1,198

- killed when the *Lusitania* was torpedoed in 1915

1,221

- letter As used in the Malaysian language for every Z used

1,225

- miles a year travelled by the average Briton commuting to work

1,253

- airports in Argentina

1,257

• female deaths from suicide and undetermined injury in UK in 1996

1,300

• average capacity of a female European brain in cubic centimetres

1,303

• accidents in UK homes in 1994 involving slippers

1,350

• population of Belgrade in 1733

1,440

• minutes in a day

1,422

• lines in the role of Hamlet

1,450

• average capacity of male European brain in cubic centimetres

1,455

• vasectomies performed in Iran, March to September 1996

1,482

• feet height of Petronas Towers, Kuala Lumpur, the world's tallest habitable building

1,486

• bastards born in Rochdale, Lancashire, 1781–1820

1,493

• passengers lost, excluding crew (out of 2,200), on the *Titanic* in 1912

1,513

• passengers and crew killed when the *Titanic* sank in 1912
• people rescued by the RAF in 1996

1,520

• officers of London's Metropolitan police off sick every day on average

1,523

• tradesmen in England and Wales prosecuted for opening on a Sunday in the year 1708–09

1,586

• daily papers in the US

1,590

• pounds of domestic waste per capita in the US each year

1,594

• fourteen-letter words in current use in English

1,595

• hours the average American watched television in 1996

1,600

• number of the White House on Pennsylvania Avenue
• references in the Bible hostile to left-handers

1,632

- miles walked each year by the average US policeman

1,638

- lectures at first World Congress of Psychotherapy in Vienna in 1996

1,652

- languages and dialects spoken in India

1,658

- metres span of the Forth Rail Bridge

1,686

- television sets per square kilometre in Singapore

1,710

- Test match runs scored by Viv Richards in 1976

1,728

- in a great gross

• cubic inches in a cubic foot
• words in the biggest vocabulary of a budgerigar

1,729

This number features in one of the best-loved mathematical anecdotes. The story goes that the great Indian mathematician Ramanujan was being visited in hospital by his friend, mentor and fellow genius, G. H. Hardy. On arriving, Hardy mentioned that he had hoped to be able to make an interesting comment about the number of the taxi he came in, but it was a very uninteresting number, 1,729. 'Oh Hardy, Oh Hardy,' Ramanujan is said to have retorted instantly, going on to point out that 1,729 is the smallest number that can be expressed as the sum of the cubes of two positive numbers in two distinct ways: $1,729 = 12^3+1^3 = 10^3+9^3$. The most difficult thing to believe about this tale is that Hardy had not also realised it immediately – unless he was just keeping quiet about it in order to cheer up his sick friend.

1,733

• aircraft lost by the Luftwaffe in the Battle of Britain

1,747

• guns seized by police in Japan in 1994

1,750

• new pet products launched in the US between 1980 and 1992

1,752

• canons in the Code of Canon Law

1,757

• cinema screens in Great Britain

1,766

• hours of sunshine in England and Wales in 1995
• miles length of the Danube, Europe's longest river

1,770

There is a beach resort called 'Seventeen Seventy' in Bustard Bay, Queensland, named after the year James Cook landed there.

• 1,770 is also the number of words in common English on which there is no general agreement on the preferred spelling (according to Lee Deighton in 1972).

1,807

• average number of people granted legal immigrant or permanent resident status in the US every day in the year 1997–98

1,813

• population of the Falkland Islands

1,815

• standard-sized graves, eight feet by three, that can be dug in an acre of land

1,821

• height in feet of the CN Tower in Toronto, the world's tallest man-made self-supporting structure

1,837

• *1837* (1951) – film which depicts a love story set at the time of the Canadian rebellion against England

1,855

• times the word 'Lord' is in the Bible

1,859

• rifles handed in to the police in Britain in 1997

1,860

• *1860* (1933) – film about a priest, a shepherd and an intellectual caught up in Sicilian revolt

1,888

• year requiring most Roman numerals under the system generally used today: MDCCCLXXXVIII. This record will not be equalled until the year 2388. The ancient Romans themselves, however, would probably not have recognised any of this, since their rules for writing numbers were different. In particular, the shorthand of writing a smaller number before a large one to indicate subtraction, such as IV for four, was introduced only in the Middle Ages. The Romans would have written four as IIII.

1,897

• miles covered by the Shah of Iran's Lamborghini Miura SVJ in 25 years

1,900

• *1900* (1976) – film starring Burt Lancaster, Donald Sutherland and Robert de Niro in twentieth-century Italian history

1,909

• letters in the full chemical name for tryptophan synthetase

1,918

• *1918* (1985) – film about a small Texas town coming to terms with World War I

1,919

• *1919* (1983) – film in which an old lady reminisces about Sigmund Freud

1,931

• *1931: Once Upon a Time in New York* (1972) – bootlegging film drama

1,936

• items of litter per kilometre of British beach found in the Beachwatch survey in 1998

1,941

• *1941* (1979) – Steven Spielberg flop with Dan Aykroyd and John Belushi in slapstick World War II farce

1,969

• *1969* (1989) – politico-comedy film drama depicting the effect of the Vietnam war on US teenagers

1,983

• kilometres the average Japanese travels on the railway each year

1,984

Title of a novel by George Orwell, filmed as:
• *1984* (1956): Edmund O'Brien is Winston Smith – withdrawn from circulation after legal problems with George Orwell's estate
• *1984* (1984): John Hurt as Winston Smith; Richard Burton also stars in his last film

1,985

• title of a novel (in homage to Orwell) by Anthony Burgess

1,993

• deaths by scorpion sting in Mexico in 1946

1,994

• men shaved in 60 minutes by Denny Rowe in Herne Bay in 1988

2,000

A popular film number:
• *2,000 Maniacs* (1964): Civil War ghost horror

• *2,000 Weeks* (1970): wife finds out about husband's mistress, Australian drama
• *2,000 Women* (1944): Dame Flora Robson in prison drama
• *2,000 Year Old Man* (1982): cartoon of Mel Brooks and Carl Reiner sketches
• *2,000 Years Later* (1969): Roman soldier awakes in modern times

2,001

• *2001: A Space Odyssey* (1968) – Kubrick's sci-fi film classic, which won an Oscar for Special Effects

2,010

• *2010: The Year We Make Contact* (1984) – the sequel to *2001*

2,020

• *2020 Texas Gladiators* (1985) – post-apocalyptic action film

2,118

• radios per thousand people in the US in 1992

2,159

• miles diameter of the moon

2,240

• pounds in a ton in the UK; the Americans settle for the simpler arithmetic of 2,000 pounds in a ton

2,251

• men killed in car crashes in the UK in 1994 (→981)

2,300

• Americans declared missing in action in the Vietnam war

2,311

• poetry books published in the UK in 1996

2,340

• meetings of the World Trade Organisation in 1996

2,343

• exclamation marks in Tom Wolfe's *Bonfire of the Vanities*

2,362

• feet length of the platform at Bournemouth railway station, the longest in the UK

2,376

• weddings in Malta in 1998

2,381

• bastards born in Rochdale, Lancashire, 1721–1820

2,391

• thefts of cars and motorcycles in Thailand from January to July 1997

2,396

• marriages in Sweden in 1990 between male doctors and female nurses (→31)

2,410

• bottles of port William Pitt the Younger is said to have drunk in a single year

2,473

• sheets of paper used by Mrs Marva Drew in typing out every number from one to a million

2,554

• percentage inflation in Angola 1995–96

2,555

• times a year an average British man fantasises about having sex

2,589

• British casualties in air raids in World War I

2,639

• manatees in Florida in 1996 – the most ever

2,714

• items of litter per kilometre of Scottish beaches found in the Beachwatch survey in 1998

2,780

• record number of baked beans eaten in 30 minutes, one by one with a cocktail stick

2,826

• four-letter words in current use in English

2,876

• wickets taken by W. G. Grace

2,895

• thirteen-letter words in current use in English

2,913

• people known to have been killed in earthquakes in 1997

2,983

• sufferers from scrofula touched by Charles II in 1669

3,000

- *The 3,000-Mile Chase* (1977) – film about drugs and gangsters

3,026

- medical incidents on British Airways flights in the year 1996–97

3,103

- prostitutes examined by the Metropolitan Police in 1837

3,106

- carats of the Cullinan diamond

3,189

- kilometres of motorway in Great Britain

3,203

- deaths from motor vehicle accidents in England and Wales in 1995

3,212

- feet drop of the Angel waterfall, Venezuela, the world's highest

3,213

• people enrolled in mortuary science programmes in the US in 1996, according to the American Board of Funeral Science Education

3,274

• miles diameter of Ganymede, moon of Jupiter, the largest moon in the solar system

3,398

• people killed on Britain's roads in 1996

3,407

• feet average depth of the Arctic Ocean

3,458

• deaths from traffic accidents in the UK in 1994

3,516

• phone calls made or received per year in the average US household

3,547

• deaths from suicide in England and Wales in 1995

3,590

• total staff at the 1997 Wimbledon Lawn Tennis Championships

3,658

• men serving jail sentences in England and Wales for sexual offences

3,708

• to 1, odds against a professional golfer making a hole-in-one on a randomly selected hole

3,712

• murders in California in 1994

3,717

• candidates in the British General Election of 1997

251 towns and cities in Vermont

3,793

- members of the American Ostrich Association in June 1995

3,798

- pairs of gloves left behind on London Transport 1997–98

3,808

- people injured in bed in the UK in 1993

3,887

- male deaths from suicide and undetermined injury in UK in 1996

3,901

- lines in *Hamlet*

3,994

- oak trees used to build Windsor Castle in the fourteenth century

4,045

- reported rapes in England and Wales in 1991

4,065

• capacity of the Metropolitan Opera, New York

4,077

• Mobile Army Service Hospital in *M*A*S*H*

4,115

• summer population of Antarctica

4,145

• length of the River Nile in miles

4,150

• average annual car mileage in the UK

4,184

• joules in a calorie

4,224

• bishops in the world according to the Vatican, February 1997

4,280

• buffaloes killed by Buffalo Bill

4,442

• handguns handed in to British police in 1997

4,613

• twelve-letter words in current use in English

4,616

• feet of the Humber Bridge, the world's longest suspension bridge

4,824

• most footnotes in an article in the *Law Review*

4,830

• British casualties in air raids in World War I

4,840

• square yards in an acre

4,870

• horse race wins by Sir Gordon Richards

4,907

• five-letter words in current use in English

4,968

• people in the world for every doctor

5,000

Films include:
• *The 5,000 Fingers of Dr T* (1953): musical fantasy with 500 child slaves forced to play the piano
• *Five Thousand Dollars on One Ace* (1964): gambling and death

5,040

• changes in a bellringer's classic eight-bell peal of 'doubles'
• ideal number of households in a city state according to Plato's Laws

5,123

• kilowatt hours of electricity consumed per capita in the UK each year

5,267

• men serving jail sentences in England and Wales for robbery

5,280

• feet in a mile

5,435

• miles of the balloon flight made by Steve Fossett in 1995 from South Korea to Canada. He broke this record in 1997 when failing in an attempt to go round the world in a balloon. In March 1999, Bertrand Piccard and Brian Jones flew 40,814 km in the first round-the-world balloon flight

5,444

• members of Skoptzy sect – who believed in self castration – on Russian police records in 1875

5,544

• area of Connecticut in square miles

5,624

• water mills recorded in the Domesday Book

5,630

- umbrellas left on LMS Railway in 1946

5,683

- new beverage brands launched in the US 1980–92

5,748

'It is two good miles, and just five thousand, seven hundred and forty-eight steps.'
Jonathan Swift, *Journal to Stella,* 1711

5,930

- reported rapes in the UK in 1996

5,938

- men serving sentences in jails in England and Wales for burglary

6,000

- *6,000 Enemies* (1939) – film with Walter Pigeon as the DA framed by the mob

6,137

• performances of *A Chorus Line* on Broadway

6,250

• people per public lavatory in Guangzhou, China

6,511

• miles travelled within Britain each year by the average Briton

6,563

• establishments in Thailand in January 1995 offering sexual services

6,704

• eleven-letter words in current use in English

6,949

• patent number of Abraham Lincoln's sole invention: 'A Device for Buoying Vessels over Shoals'

7,000

• number of people meditating simultaneously that it takes to influence world affairs according to Maharishi Mahesh Yogi

7,254

• *7254* (1971) – film: war drama

7,612

• parking tickets issued to the Soviet UN mission in 1989

7,794

• murders in Rio de Janeiro in 1997

7,910

• six-letter words in current use in English

7,985

• road fatalities in the UK in 1966, the worst year on record

8,280

• miles from Chicago to Hong Kong on the world's longest nonstop air trip, with United Airlines, inaugurated in July 1996

8,515

• men serving jail sentences in England and Wales in 1997 for violence against the person

8,771

• women arrested for prostitution in England and Wales in the year to 29 September 1857

8,833

• horse race winners ridden by Willie Shoemaker

9,000

• the weight in grams of 9,000 metres of yarn is equal to its number of denier

9,091

• people per doctor in Swaziland – the world's worst ratio

9,153

• ten-letter words in current use in English

9,353

• entries in the first Chinese dictionary, compiled in AD 100

9,665

• shoe shops in Great Britain

9,906

• seven-letter words in current use in English

10,000

Films:
• *Ten Thousand Bedrooms* (1957): musical comedy with Dean Martin
• *10,000 Dollars Blood Money* (1966): Italian film about a bounty hunter

10,080

• umbrellas left on London Transport 1997–98

10,404

- dispensing chemists in Great Britain

10,442

- different words in the King James Bible, according to Lincoln Bennett

10,791

- nine-letter words in current use in English

11,033

- eight-letter words in current use in English

11,174

- Test runs scored by Allan Border

11,610

- words in the role of Hamlet

11,738

- camels in Australia in 1922

12,139

• Americans who bet on General Noriega's prison number in the first Florida State Lottery after his arrest

12,634

• butchers' shops in Great Britain

12,769

• people killed by handguns in the US in 1994

13,240

• penises cut off as trophies of the ancient Egyptian victory of Pharaoh Meneptah over the Libyans at Karnak

13,272

• sex workers in Bangkok

13,387

• airports in the US

13,759

• miles along the frontier of China

13,797

• prison inmates in the UK who tested positive for drugs between 1997 and 1999

14,299

• pounds of jellies, pickles and hams sold at the Great Exhibition of 1851

15,487

• pages in the first completed revised edition of the *Oxford English Dictionary*

15,502

• people per square mile in San Francisco

15,771

• height of Mont Blanc in feet

16,091

• deaths from injury and poisoning in the UK in 1996

16,136

• gun murders in the US in 1993

17,561

• kilometres of railway in the UK

17,677

• different words used by Shakespeare

17,968

• kilometres of the coastline of Antarctica

18,033

• traffic accidents on the roads of Morocco in the first six months of 1996

19,711

• dentists in the UK

20,000

A popular number in films:
• *20,000 Leagues Under the Sea* (1916): silent classic
• *20,000 Years in Sing-Sing* (1933): Spencer Tracy as a reformed criminal in love with Bette Davis
• *20,000 Men a Year* (1939): adventure thriller with Randolph Scott
• *20,000 Leagues Under the Sea* (1954): Kirk Douglas and James Mason in Disney's version of Jules Verne, which won an Oscar for Special Effects
• *20,000 Eyes* (1961): embezzler steals from gangster suspense
• *The 20,000-Pound Kiss* (1963): Edgar Wallace mystery with Dawn Addams

20,352

• people killed by scorpions in Mexico 1940–49

21,785

• words in the *OED* that had their first appearance in the eighteenth century

23,305

• murders in the US in 1994

23,530

• gallstones removed from an eighty-five-year-old woman in Worthing, Sussex in 1987

23,865

• people per square kilometre in Macao, the world's most densely populated place

25,550

• bottles of suntan lotion bought by the US Army from one K-Mart store in Georgia in the month after Iraq invaded Kuwait

26,963

• kilograms of milk in 377 days mostly in 1995 produced by Acme Gold 2nd at Kettering, Northamptonshire – a new world record for one (slightly extended) year's production

27,102

• suicides in Japan in the first ten months of 1998

27,457

• times Shakespeare used the word 'the'

29,022

• height of Everest in feet

29,899

• different words used by James Joyce in *Ulysses*

30,107

• clothes shops in Great Britain

30,135

• police officers in New York city

30,510

• square kilometres in Belgium

31,388

• parking violation tickets ignored by the Russian UN mission in New York in 1996

32,354

• words listed in the *OED* that were already in the language by 1400

32,429

• labrador retrievers registered by the Kennel Club – making it the UK's most popular pedigree breed

32,738

• homicides in Colombia in 1996

35,797

• feet depth of Marianas trench in Pacific – the deepest spot in the ocean

35,810

• words in the *OED* that had their first appearance in the 16th century

40,000

• *Forty Thousand Horsemen* (1941) – war film with Chips Rafferty

40,111

• cars stolen in Rio de Janeiro in 1997

41,522

• traffic accidents on roads in Morocco in 1995 – the country with the world's worst accident rate per car

41,530

• Sherman tanks produced in World War II

42,551

• bombs dropped on Japan in July 1945

43,560

• square feet in one acre

44,473

• people seriously injured on Britain's roads in 1996

46,227

• times the word 'and' appears in the Bible

46,773

• words in the *OED* that had their first appearance in the seventeenth century

50,000

Just two films to note:
• *Fifty Thousand Dollar Reward* (1924): silent western
• *50,000 BC (Before Clothing)* (1963): sex comedy

51,047

• average daily prison population in England and Wales in 1995

52,415

• people per square mile in Manhattan

54,393

• people in British jails in May 1996

54,589

• deaths on the road in the US in 1972, the record number for a calendar year

54,659

• Americans injured by ovens and stoves in 1991

54,896

• runs made by W. G. Grace

59,211

• cost in pounds of the wallpaper in the Lord Chancellor's official residence

59,328

• motorcycles sold in the UK in 1996

59,905

• words in this book including the bibliography, and index

60,559

• words in the *OED* that had their first appearance in the nineteenth century

63,360

• inches in a mile

64,747

• dollars in each US family's share of the national debt

68,578

• deaths in London caused by the great plague of 1664–65 according to official figures

70,000

• *70,000 Witnesses* (1932) – film in which someone is murdered at a football game

71,280

• prostitutes in Indonesia in 1994, according to official figures

71,852

• possible positions in a chess game after Black's second move

73,884

• people killed by the atom bomb dropped on Nagasaki, 9 August 1945

75,981

• the chance of being murdered within the next year for a UK resident is 1 in 75,981

80,000

• *80,000 Suspects* (1963) – film with Claire Bloom in a tale of smallpox in Bath

86,400

• seconds in a day

88,888

• account number for Nick Leeson's illicit Barings operations in Singapore

100,000

• when the Ukraine changed its currency from the karbovanets to the hryvnia in 1996, the exchange rate was 100,000 karbovantsi to 1 hryvnia

101,504

• books published in the UK in 1996

101,683

• convictions secured by the Societies for Promoting a Reformation of Manners for offences against public decency between 1698 and 1738

106,307

• seminarians studying philosophy and theology, according to Vatican in February 1997

121,697

• patents granted in the US in 1996

131,153

• complaints to local authorities in the UK about noisy neighbours in 1993–94

134,281

• parking tickets ignored by UN diplomats in New York in 1996

142,807

• people killed in the Tokyo-Yokohama earthquake in 1923

144,000

• days in a 'baktun' – 400 years of the Mayan calendar, each year comprising 18 weeks each of 20 days

149,547

• spectators at the Scotland–England football match at Hampden Park in 1937

151,485

• miles length of the coastline of Canada

155,499

• divorces in England and Wales in 1995

174,465

• patent number for Bell's telephone, the most valuable patent in history

177,737

• the world record for pogo stick jumps

186,272

• speed of light in miles per second

187,880

• lakes in Finland

201,148

• outlets for buying alcohol in the UK

211,208

• square miles of France

223,898

• US patent number for Edison's electric light

289,996

• shops in Great Britain according to the latest (1997) statistics

308,982

• male deaths in the UK in 1995

322,751

• items of litter found on 257 British beaches in the Beachwatch Survey in 1998

324,198

• stars catalogued by Friedrich Argelander in his *Bonner Durchmusterung*, 1859–62

331,248

• marriages in the UK in 1994

332,730

• female deaths in the UK in 1995

338,876

• cigarette butts picked up in a one-day coastal clean-up in California in September 1998

366,999

• kilometres of public roads in Great Britain

398,671

• British troops killed in the first battle of the Somme in 1916

404,750

• priests on earth according to the Vatican in February 1997

414,825

• entries in the *OED* (first complete revised edition, 1933)

450,000

• cubic cubits volume of Noah's Arc (300 cubits long by 50 wide and 30 high)

525,539

• cars sold in the UK in August 1997, the highest ever monthly total

529,000

• *The $529,000 Boo-Boo* (1971) – film in which a hermit is accidentally credited with a large sum of money by his bank

565,939

• pills taken by C. H. A. Kilner of Zimbabwe between 1967 and 1988

640,145

• number of the US population who were born in England (1990 census)

658,958

• flights made by UK airlines in 1995

773,696

• words in the King James Bible

802,701

• Final year arrived at by the traveller in H. G. Wells's *The Time Machine*

870,027

• plain buns sold at the Great Exhibition of 1851 in London

934,691

• Bath buns sold at the Great Exhibition of 1851 in London

967,500

• square miles area of Sudan (the largest country in Africa)

1,000,000

When the word 'million' occurs in a film title, there is a greater than one-in-three chance that the next word will be 'dollar'.

Here are some 'million' films:

• *If I Had a Million* (1932): George Raft and W. C. Fields in a crime comedy

• *Million Dollar Legs* (1932): comedy with Ben Turpin and W. C. Fields
• *Million Dollar Ransom* (1934): based on a story by Damon Runyon *One Million Dollars Ransom*
• *I Stole a Million* (1939): George Raft and Jason Robards in crime mystery
• *Tanks a Million* (1941): war comedy
• *Million-Dollar Baby* (1941): comedy drama with Ronald Reagan
• *The Missing Million* (1942): based on a book by Edgar Wallace
• *A Girl in a Million* (1946): Michael Hordern and Joan Greenwood in a comedy
• *Million-Dollar Mermaid* (1952): Walter Pigeon and Victor Mature in a swimming spectacular
• *The Beast With a Million Eyes* (1956): sci-fi horror western
• *Million-Dollar Collar* (1964): a tale of a performing dog and jewel smugglers
• *How to Steal a Million* (1966): Audrey Hepburn hires Peter O'Toole to do a robbery
• *One Million Years BC* (1967): prehistoric anachronisms with Raquel Welch
• *The Million Eyes of Su-Muru* (1967): spy story with Klaus Kinski and Shirley Eaton
• *Million-Dollar Duck* (1971): Disney fantasy with golden eggs
• *A Million to Juan* (1994): based on Mark Twain's story: *The Million-Dollar Banknote*

1,196,279

• children aged 10–16 needing treatment for sports injuries in the UK in 1997

1,226,467

• people saw the Tutankhamun exhibition at Metropolitan Museum in New York 1978–79

1,238,085

• pounds taken on the opening day of *Batman Forever* in the UK

1,694,117

• attendance at the Tutankhamun exhibition at the British Museum 1972–73

2,000,000

A good number for eccentric films:
• *The Two Million Clams of Cap'n Jack* (1973): George Peppard drama
• *Cop Gives Waitress $2 Million Tip* (1994): romantic comedy with Bridget Fonda

2,054,754

• pounds of the largest unclaimed prize in the British National Lottery

2,102,338

• overnight stays in youth hostels in England and Wales in 1996

2,367,234

• Malaysian 20-sen coins used to set a world record in 1996 for the longest line of coins. The mark to beat is now 55.63 kilometres.

3,586,489

• letters in the Bible

4,000,000

• 'The Four Million' – short story by O. Henry

5,000,000

• *Five Million Years to Earth* (1968) – sci-fi horror film

5,118,470

• pairs of green socks bought for US armed forces in 1989

5,195,930

• Britons who emigrated to the US between 1820 and 1994

5,506,720

• documents classified as secret or top secret by the US government in 1989

6,000,000

• *Cyborg: The Six-Million Dollar Man* (1973) – film with Lee Majors

6,469,952

• spots drawn by animators for the Disney film of *101 Dalmatians*

7,126,132

• Germans who emigrated to the US between 1820 and 1994

8,000,000

• *8 Million Ways to Die* (1985) – film starring Jeff Bridges in a cops, pimps and prostitutes drama

8,765,832

• hours from the start of 1 January 2,000 until the end of 31 December 2,999

10,000,000

• ergs in a joule

Film:
• *The Ten Million Dollar Getaway* (1991): made-for-TV drama of a 1978 robbery at Kennedy Airport

11,914,200

• new car registrations in Western Europe in 1994

13,247,091

• square miles in the British Commonwealth at the beginning of 1997

20,000,000

Films:
• *20 Million Miles to Earth* (1957): US rocket to Venus brings back rampaging alien
• *20 Million Sweethearts* (1934): Ginger Rogers and Dick Powell in a musical comedy

50,000,000

• *Fifty Million Frenchmen* (1931) – Cole Porter musical film comedy

53,310,761

• Elvis Presley's army serial number

59,113,439

• population of the UK (July 1999 estimate)

107,689,927

• animals killed by the US fur trade between 1919 and 1921

148,081,443

• cars in the US (which works out at 1.8 persons per car)

266,476,278

• population of the US in July 1996

399,902,004

• angels in the universe: in nine choirs of 6,666 legions, each legion having 6,666 spirits according to mediaeval theologians

500,000,000

• *Objective 500 Million* (1966) – French-Italian co-production on a film about a plot to steal 500m francs

1,000,000,000

In the 16th century, the English adopted the term 'billion' to mean a million million. This apparently led to no confusion until the late eighteenth century when the French started generally using the word 'billion' to mean a thousand million. Indeed, there is evidence of 'byllion' and 'tryllion' having been used in France as early as the 15th century to mean 1000 million and a million million respectively. American usage in the late 18th century followed the French version. For the first half of the twentieth century there was confusion between American and British billions, but since the 1950s, the British have increasingly followed the Franco-American lead. The correct prefix for a billion of anything is 'giga-'.

A billion, as seen in the films:

• *The Billion Dollar Hobo* (1932): man has to live as tramp to collect inheritance

• *Billion Dollar Brain* (1967): Ken Russell directs Michael Caine in the Len Deighton story

• *Mr Billion* (1977): Jackie Gleason comedy adventure

• *A Billion for Boris* (1990): man rewires television to pick up programmes from the future

9,000,000,000

• 'The Nine Billion Names of God' – short story by Arthur C. Clarke

9,192,631,770

• cycles of resonance vibration of the caesium-133 atom in a second. Since 1967 this has been the formal scientific definition of a second

11,424,596,055

• pounds spent on the UK National Lottery in its first three years

15,654,023,458

• inches from the earth to the moon as measured at 16.425145 seconds past 3.42 p.m. on 24 November 1997

4,985,567,071,200

• dollars of US national debt when the clock stopped near Times Square, New York, on 14 November 1995

5,671,350,435,227

• dollars of US national debt two weeks before Christmas 1999

APPENDIX: Percentage Pointlessness

As mentioned earlier, we have restricted the number of percentages included in the main body of this work, firstly because percentages are not true, countable numbers, and secondly because they often appear as a consequence of surveys of dubious methodology. For anyone requiring a different sample of percentages from those found under 50, this appendix collates selected results from a large number of surveys conducted in 1999. All samples, unless otherwise stated, are British adult populations.

1 per cent of

- women would choose Robin Cook if they had to pick a politician to cuddle up to
- British tourists take English teabags on overseas trips
- men aged 21–30 think lollipop ladies are the sexiest women in uniform

2 per cent of

- women would choose William Hague if they had to pick a politician to cuddle up to
- adults don't buy any Christmas presents

3 per cent of

- men spend more than £2,000 a year on romantic gestures
- house-buyers think it is important to live near a cinema

4 per cent of

- prisoners serving life sentences are women
- the population spend less than one hour a week cleaning the house
- men aged 21–30 think nuns are the sexiest women in uniform

5 per cent of

- men are afraid of domestic violence
- house-buyers say that easy access to fast food is important to them

6 per cent of
- women feel less interested in sex after strenuous sessions in the gym
- 16–24-year-olds say they take illegal drugs daily

7 per cent of
- Scots say they have split up with a partner on holiday
- men aged 16–25 believe that bringing up children is women's work

8 per cent of
- women over 16 are divorced
- 13–16-year-olds claim to have been bullied in the past six months

9 per cent of
- couples say relatives are the main cause of tension in their relationship
- adults admit to having had unsafe sex with a new partner

10 per cent of
- men feel let down by the thrill of a one-night stand
- US women have ended a romance because their man didn't like their cat

11 per cent of
- women would choose Tony Blair if they had to pick a politician to cuddle up to
- working women work more than 51 hours a week
- men think their wife or girlfriend is the funniest person they know

12 per cent of
- the fast-food market is fish and chips
- American women would like to change the size of their breasts

13 per cent of
- people are willing to travel more than 10 miles to go out for a meal
- couples say housework is the main cause of tension in their relationship

14 per cent of
- single men aged 28–33 want to stay single for ever
- adults say they would be willing to consider having sex in a taxi

15 per cent of
- accountants never drink
- accountants smoke
- people think journalists tell the truth

16 per cent of
- women do not feel safe in their own home
- adults have placed bets in betting shops in the last two years

17 per cent of
- couples say their sex lives get worse the longer they stay together
- our average household expenditure goes on food

18 per cent of
- people would trust Cliff Richard to sell them a decent pension
- the fast-food market is burgers

19 per cent of
- the Welsh can speak Welsh
- people can't remember when they last said 'I love you'

20 per cent of
- men would agree to their child taking the mother's surname
- women think their husband or boyfriend is the funniest person they know

21 per cent of
- families are headed by a single parent
- women drivers lose confidence when they have a male passenger

22 per cent of
- men think it is important to live near a supermarket
- all couples say money is the main cause of stress

23 per cent of
- women say they have suffered domestic violence at some time
- married women don't think it's important to have their own money
- people think politicians and government ministers tell the truth

24 per cent of
- Americans think Germany is the most influential country in Europe
- gardeners sometimes use chemical weed-killers

25 per cent of
- public toilets on Britain's beaches are badly maintained
- women say they would wear no underwear for a very special date

26 per cent of
- men say they would give up smoking to please a partner
- parents say that what they most want for their children is wealth

27 per cent of
- full-time employees in the UK work more than 46 hours a week
- 7–12-year-olds claim to have been bullied in the last 6 months

28 per cent of
- women smoke
- workers say their office suffers from bad ventilation

29 per cent of
- *Cosmopolitan* readers say they like the missionary position for sex best
- adults say they would be willing to consider having sex in the office
- men smoke

30 per cent of
• adults would consider an abortion if their unborn baby had a serious inherited disease

31 per cent of
• female cat-owners would rather their partner went missing than their cat

32 per cent of
• men said they intended to spend nothing on Valentine's Day
• women have been victims of vandalism

33 per cent of
• women wish their breasts were firmer
• American parents think a return to school prayers would cut violence

34 per cent of
• the English own their own homes
• people take their umbrella away with them

35 per cent of
• men feel jealous if they see their girlfriend chatting to another man
• women think it is important to live near a supermarket

36 per cent of
• adults admit to having fantasised about someone from their past when having sex

37 per cent of
• UK men say their kissing technique leaves a lot to be desired
• women say buying things for the children is the main temptation to overspend
• people think health is more important than happiness

38 per cent of
- people think the police are racist
- male smokers say cigarettes are their main source of pleasure

39 per cent of
- under-44s claim to have made love on a foreign beach
- parents are worried about teenage pregnancy

40 per cent of
- the world population are under 15
- men who live alone would prefer to be with a partner
- hereditary peers are bastards or descended from bastards

41 per cent of
- the fast-food market is sandwiches

42 per cent of
- 14–16-year-old girls felt Dad was a better driver than Mum
- women believe that women are natural born house cleaners
- teenagers say they have lost sleep through worrying about exams
- smokers say they are not concerned about the health effects

43 per cent of
- drivers said they would use their car less if public transport were better
- people find flat-pack furniture assembly instructions a mystery

44 per cent of
- women smokers say cigarettes are their main source of pleasure
- parents say what they most want for their children is 'success'
- men under 35 love cooking

45 per cent of
- Belgian footballers say their club is not managed properly

46 per cent of

• teenage boys wear a helmet when cycling
• women think paedophiles should be castrated

47 per cent of

• parents have babies that regurgitate after every feed
• Londoners don't know their neighbours

48 per cent of

• 14–16-year-olds think they can earn a living from their favourite pastime

49 per cent of

• people are often too busy to spend quality time with their nearest and dearest
• mothers are worried that their child might be abducted

50 per cent of

• adults want to be taught 'parenting skills' when they have a child
• British tourists stick to fish and chips and English breakfasts abroad
• all journeys are less than two miles

51 per cent of

• leisure travellers take short breaks rather than longer holidays because of work pressures

52 per cent of

• men said they were first attracted by their future partner's figure
• teenagers are worried about the environment

53 per cent of

• women dislike the house they live in
• fathers have consulted their mother-in-law for advice on parenting

54 per cent of

• Swedish births are outside marriage

• men aged 16–34 are concerned about cancer and heart disease

55 per cent of

• the French find German tourists agreeable

• men shy away if someone they don't know well tries to greet them with a kiss

56 per cent of

• Londoners think the Millennium Dome is a waste of money

• cat-owners buy their pets a present at Christmas

• men would love to wear a suit of armour

57 per cent of

• women think rapists should be castrated

• teenage girls wear a helmet when cycling

58 per cent of

• 14–16-year-old boys felt Dad was a better driver than Mum

• male dog-owners say they often find it easier to talk to their dog than to their partner

59 per cent of

• people think Prince Charles should still be king if he marries Camilla

60 per cent of

• people believe that the ordinary man in the street tells the truth

• Americans think Britain is the most influential country in Europe

• doctors believe stroking pets can lower blood pressure and heart rate

61 per cent of

• people don't know in which century the English civil war took place

62 per cent of
• teenagers are worried about homelessness
• dog-owners say their dog can sense their moods

63 per cent of
• women think it is sexist for a child to take its father's surname

64 per cent of
• doctors don't wash their hands properly
• parents are worried about the issues of drug and alcohol abuse

65 per cent of
• people find reading enjoyable
• music fans believe that rock singers help to incite drug use

66 per cent of
• people say the police are too remote from the public
• 16-year-olds say they are pessimists

67 per cent of
• Americans aged between 18 and 34 say they would stop to pick up a penny lying on the ground
• men aged 16–25 would not get married just because their girlfriend became pregnant

68 per cent of
• British houses have at least one smoker
• women would love to be a Victorian lady of leisure
• police officers in England and Wales said they would take another professional job with the same rate of pay if it were offered to them

69 per cent of
• people always read to the end of any novel they start

70 per cent of
• men use aftershave every day
• dog-owners buy their pets a present at Christmas

71 per cent of
• first-time mothers feel it is socially frowned upon to choose to be a full-time mother
• women are in favour of hanging for murderers

72 per cent of
• families buy takeaways because they haven't time to cook in a typical week

73 per cent of
• families with children think their lives are becoming more difficult
• London men say they have had sex with a colleague at work
• women do all the ironing

74 per cent of
• parents think it's all right to smack a child's hands, legs or bottom
• people mistrust journalists
• pub-goers approve of no-smoking zones

75 per cent of
• women aged 25–44 are either in work or seeking it
• first-time brides receive a new diamond engagement ring

76 per cent of
• pets get a slice of the Christmas turkey on 25 December
• people say Camilla Parker-Bowles should never become Queen

77 per cent of
• working women say they would quit if they could
• people use different chopping boards for raw meat and other products

78 per cent of

• women wear pyjamas or night-gowns in bed in hotels
• women believe that marriage is vital for stable family life

79 per cent of

• the public say they don't believe what they read in the papers
• people most enjoy reading in the living room

80 per cent of

• women would not have breast implants even if they were free
• Scottish women are prepared to spend the rest of their lives with their current partner

81 per cent of

• people have faith in teachers
• people believe space travel will be an option in the third millennium
• people believe that marriage and relationship problems can trigger depression

82 per cent of

• adults wear aftershave or perfume every day

83 per cent of

• drivers say they would find it very difficult to adjust their lifestyle to being without a car
• people think the police are mainly helpful and polite
• dog-owners say their dog is as much a part of the household as the human members

84 per cent of

• Britons living in the US prefer the UK's public transport system
• working women think that women perform too many roles nowadays
• children expelled from school are boys

85 per cent of

• married men between 28 and 33 don't regret getting married

• teenage girls think they should know a man for at least a year before marrying

86 per cent of

• men aged 16–25 say they would accept financial responsibility for their children

87 per cent of

• British houses have a dog, cat or smoker

• pet-owners spend over £5 on Christmas presents for their dog or cat

88 per cent of

• politicians are aware of the pay gap between men and women

• people see financial problems as a trigger for depression

89 per cent of

• cat- or dog-owners send the pet out of the bedroom before they have sex

90 per cent of

• Americans over 65 say they would stop to pick up a penny if they saw it on the ground

• lung-cancer cases are caused by smoking

91 per cent of

• women complain that their partner never helps clean the kitchen and bathroom

• people think we will one day live and work in space

92 per cent of

• people think the Noddy books have been good for children

93 per cent of

- adults are aware that cholesterol levels may be a health issue
- people believe that anyone who can't afford legal bills should receive aid
- mothers say that becoming a parent for the first time was the most joyful experience of their lives

94 per cent of

- health authorities have experienced recruitment problems
- women drivers think men need to improve their attitude towards them
- pet-owners describe their pet as 'just another member of the family'

95 per cent of

- US young adults say kissing in public is totally stimulating
- people like to receive books as presents
- people think reading is an important and positive part of their lives

96 per cent of

- Scots believe they have a responsibility to keep themselves healthy
- Spaniards are cremated
- people sometimes give books as presents

97 per cent of

- gardeners prefer to weed by hand or hoe and not use chemicals
- olive oil is accurately described on its label

98 per cent of

- Viagra prescriptions in the US are to men over 40

99 per cent of

- people don't know the fat content of milk

100 per cent of

- surveys in popular magazines ought not to be taken too seriously

Bibliography

Innumerable works of reference have been consulted in the compilation of this book. The following are some of the most useful and intriguing.

Works on numbers:

• *The Guinness Book of Numbers* (1989), Adrian Room – a useful general work on the history of numbers, with a good deal of quirky information

• *The Mystery of Numbers* (1993), Annemarie Schimmel – all the mysticism and religious symbolism of numbers

• *The Penguin Dictionary of Curious and Interesting Numbers* (1986), David Wells – everything you want to know about the mathematical side of numbers

General reference works on CD-ROM:

• *The Oxford English Dictionary* – generally viewed as a collection of words, this fine work is also very rich in numbers

• *The Oxford Interactive Encyclopedia*

• *The World Book Multimedia Encyclopedia*

• *The Hutchinson Multimedia Encyclopedia*

• *The New Grolier Multimedia Encyclopedia* – the search facility on all these CD-ROMs enables one to look up numbers in a way that was impossible with traditional alphabetical ordering

• *The Complete Works of Jane Austen* – not renowned for her use of numbers, this authoress does, however, include some useful observations on 19, 48, 100, 130 and the size of dinner-parties

• *The Complete Works of William Shakespeare* – a writer whose skill with words has been allowed unfairly to overshadow his general numeracy

• *The Bible* – a rich source of numerical references, although its own 'Book of Numbers' is disappointing in this respect

Subsidiary sources:

• *Concise Oxford Dictionary of the Christian Church* (1977), ed. Elizabeth Livingstone – contains a few useful numbers

• *All the Trouble in the World* (1994), P. J. O'Rourke – full of useful statistics

• *The Black Arts* (1967), Richard Cavendish – tips on numerology and alchemy

• *The Body* (1985), Anthony Smith – contains many more numbers than the same author's *The Mind*
• *Brewer's Dictionary of Phrase and Fable* (1894) – the first edition of this classic is by far the most numerate
• *Cluck!* (1981), Jon-Stephen Fink – the definitive work on chickens in films
• *The Dent Dictionary of Measurement* (1994), Mike Darton and John Clark – very useful
• *Human Development Report* (1966), United Nations Development Programme – packed with useful figures
• *Innumeracy* (1989), John Allen Paulos – with fewer numbers than one might expect, but those that appear are generally well chosen
• *The Larousse Dictionary of Science and Technology* (1955) – more words than numbers, but still informative
• *Made in America* (1994), Bill Bryson – a brilliantly informative collection of words and numbers
• *The Medieval Machine* (1988), Jean Gimpel – excellent on ancient numbers
• *The Mother Tongue* (1990), Bill Bryson – a perfect balance between numerical and verbal information
• *The New Penguin Dictionary of Music* (1986) – a handy guide apart from its inconsistency on the number of Hungarian Rhapsodies written by Liszt
• *Ostrich Egg-shell Cups of Mesopotamia and the Ostrich in Ancient and Modern Times* (1926), Berthold Laufer – a classic
• *Phrenology* (1969), Orson Squire Fowler and Lorenzo Niles Fowler – a modern reprint of a nineteenth-century classic on head bumps
• *Population Trends* (1996-99), Office for National Statistics – full of numbers, though rather too many are rounded to the nearest thousand
• *Sex in Georgian England* (1994), A. D. Harvey – definitive
• *Sports Spectators* (1986), Allen Gutman – an indispensable history of non-participation
• *The Total Package* (1995), Thomas Hine – all you need to know about cans, boxes and tubes
• *The Top Ten of Everything* (1997–99), Russell Ash – the thinking man's book of records
• *What Counts – The Complete Harper's Index* (1991), ed. Charis Conn and Ilena Silverman – a fine collection of figures, sadly not arranged in numerical order

Index

Items or topics counted, measured or otherwise enumerated in the text (references are to the numbers under which relevant items appear)